Lecture Notes in Mathematics 1480

Editors:
A. Dold, Heidelberg
B. Eckmann, Zürich
F. Takens, Groningen

Subseries:
Instituto de Mathemática Pura e Aplicada
Rio de Janeiro, Brazil (vol. 48)

Adviser:
C. Camacho

F. Dumortier R. Roussarie
J. Sotomayor H. Żołądek

Bifurcations of Planar Vector Fields

Nilpotent Singularities and Abelian Integrals

Springer-Verlag
Berlin Heidelberg New York
London Paris Tokyo
Hong Kong Barcelona
Budapest

Authors

Freddy Dumortier
Limburgs Universitaire Centrum
Universitaire Campus
3610 Diepenbeck, Belgium

Robert Roussarie
Département de Mathématiques
Université de Bourgogne
UFR de Sciences et Techniques
Laboratoire de Topologie
(U. A. no. 755 du CNRS), B. P. 138
21004 Dijon, France

Jorge Sotomayor
Instituto de Matemática Pura e Aplicada
Estrada Dona Castorina 110
CEP 22460 Jardim Botânico
Rio de Janeiro, Brazil

Henryk Żoładek
Institute of Mathematics
Warsaw University
00-901 Warsaw, Poland

Mathematics Subject Classification (1991): 58F14, 34C05, 34D30

ISBN 3-540-54521-2 Springer-Verlag Berlin Heidelberg New York
ISBN 0-387-54521-2 Springer-Verlag New York Berlin Heidelberg

Printed in Germany

Typesetting: Camera ready by author
Printing and binding: Druckhaus Beltz, Hemsbach/Bergstr.
46/3140-543210 - Printed on acid-free paper

PREFACE

The study of bifurcations of families of dynamical systems defined by vector fields (i.e. ordinary differential equations) depending on real parameters is at present an active area of theoretical and applied research.

Problems in mathematical biology, fluid dynamics, electrical engineering, among other applied disciplines, lead to multiparametric vector fields whose bifurcation analysis of equilibria (singular points) and oscillations (cycles) is required.

The case of planar vector fields, due to the presence of regular as well as singular limit cycles is the first one, in increasing dimension of phase space, whose study cannot be fully reduced to the analysis of singularities and zeroes of algebraic equations, particularly when the number of parameters involved is larger than or equal to two.

The results established in this volume illustrate the diversity of the algebraic, geometric and analytic methods used in the description of the variety of structural patterns that appear in the bifurcation diagrams of generic three-parameter families of planar vector fields, around singular points whose linear parts are nilpotent.

The analysis involved in their proofs and in the discussion of the remaining conjectures points out to the actual limits of established tools for the study of complex bifurcation problems.

The introductions to the two works which constitute this volume locate precisely, in the context of the current literature, the specific character of each of their contributions.

The authors

Generic 3-Parameter Families of Planar Vector Fields, Unfoldings of Saddle, Focus and Elliptic Singularities With Nilpotent Linear Parts

by

F. Dumortier, R. Roussarie, J. Sotomayor

Contents of the volume

Table of contents

PART I: PRESENTATION OF THE RESULTS AND NORMALIZATION

PART II: RESCALINGS AND ANALYTIC TREATMENT

PART I : PRESENTATION OF THE RESULTS AND NORMALIZATION

CHAPTER I : INTRODUCTION

I.1 : Position of the problem and statement of results

This paper continues the study of 3-parameter families of vector fields in the plane, initiated in [DRS]. We present here three unfoldings of singularities of codimension 3. These unfoldings, together with those mentioned or studied in [DRS], provide a complete list of topological models for all the possible generic local k-parameter families for $k \le 3$. The reader is referred to the introduction of [DRS] for a review of known facts concerning unfoldings used freely here.

We consider germs of vector fields (at $0 \in \mathbb{R}^2$) with nilpotent 1-jet; that is with a linear part linearly conjugate to $y \frac{\partial}{\partial x}$. Such a germ has a 2-jet C^∞-conjugate to :

$$y \frac{\partial}{\partial x} + (\alpha x^2 + \beta xy) \frac{\partial}{\partial y} \qquad (1)$$

(For the definition of conjugacy, see Chapter II).

In [DRS] we studied the <u>cusp case</u> corresponding to $\alpha \neq 0$ and $\beta = 0$. Here, we study the other possibilities of codimension 3, when $\alpha = 0$ and $\beta \neq 0$. The germs of vector fields at $0 \in \mathbb{R}^2$, whose 1-jets are nilpotent form a manifold of codimension 2 in the space of all singular germs. Therefore the 2 sets of conditions : $\alpha \neq 0$, $\beta = 0$ and $\alpha = 0$, $\beta \neq 0$ define 2 sets of codimension 3. We want to study 3-parameter local families which generically unfold a germ belonging to a regular part of these 2 sets, defined by removing a subset of codimension 4.

For the cusp case ($\alpha \neq 0$, $\beta = 0$) we defined in [DRS] the regular part to be the subset of germs whose 4-jet is C^∞-equivalent to :

$$y \frac{\partial}{\partial x} + (x^2 \pm x^3 y) \frac{\partial}{\partial y} \qquad (2)$$

This set is the union of 2 submanifolds Σ^3_{C+} and Σ^3_{C-}, obtained by removing the subset of germs whose 4-jet is equivalent to $y \frac{\partial}{\partial x} + x^2 \frac{\partial}{\partial y}$.

We know by [T] that the germs in the second set ($\alpha = 0$, $\beta \neq 0$) have a 4-jet C^∞ conjugate to :

$$y \frac{\partial}{\partial x} + (\epsilon_1 x^3 + dx^4 + bxy + ax^2 y + ex^3 y)\frac{\partial}{\partial y} \tag{3}$$

with $b>0$, $\epsilon_1 = 0, \pm 1$.

It was shown in [D] that the topological type of such a germ is determined by its 3-jet, if $\epsilon_1 \neq 0$ and $b \neq 2\sqrt{2}$ in case $\epsilon_1 = -1$. The regular part of our set is defined by these conditions and the extra condition

$$5\,\epsilon_1\, a - 3b\, d \neq 0 \tag{3'}$$

which we will explain in Chapter III, third step. We will also show that the 4-jet is then C^∞-equivalent to :

$$y \frac{\partial}{\partial x} + (\epsilon_1\, x^3 + bxy + \epsilon_2 x^2 y + f\, x^3 y)\frac{\partial}{\partial y} \text{ with } \epsilon_{1,2} = \pm 1,\ b > 0. \tag{3"}$$

The topological type falls into one of the following categories :

1) <u>The saddle case</u> : $\epsilon_1 = 1$, any ϵ_2 and b. (a degenerate saddle). Here we make a distinction according to the sign of ϵ_2.
We denote by $\Sigma^3_{S\pm}$ ($\epsilon_2 = \pm 1$), the subsets of germs with such a 4-jet.
They all have the same topological type.

2) <u>The focus case</u> : $\epsilon_1 = -1$ and $0<b<2\sqrt{2}$. (a degenerate focus). We denote by $\Sigma^3_{F\pm}$ ($\epsilon_2 = \pm 1$) the corresponding subsets of germs.

3) <u>The elliptic case</u> : $\epsilon_1 = -1$ and $b>2\sqrt{2}$. Notation $\Sigma^3_{E\pm}$ with $\epsilon_2 = \pm 1$.
They all have the same topological type.

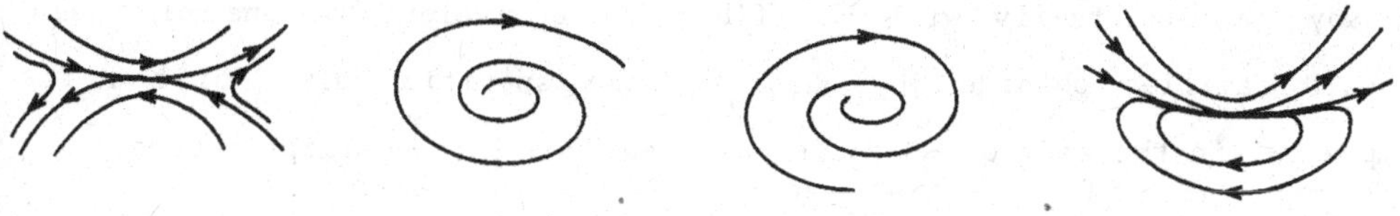

$\epsilon_2=1$ $\qquad$ $\epsilon_2=-1$

Saddle case $\qquad$ Focus cases $\qquad$ Elliptic case

Fig. 1. : The different topological types

The union of the 6 subsets $\Sigma^3_{S\pm}$, $\Sigma^3_{F\pm}$, $\Sigma^3_{E\pm}$ is the regular part of our second set. Each of them is a submanifold of codimension 3 in the space of singular germs of vector fields (i.e. which vanish at $0 \in \mathbb{R}^2$) and so, a codimension 5 submanifold of the space of all germs. We want to study generic local 3-parameter families X_λ, with $X_0 \in \Sigma^3_{S\pm}$, $\Sigma^3_{F\pm}$ or $\Sigma^3_{E\pm}$. $(\lambda \in \mathbb{R}^3)$.
The genericity condition consists in the transversality of the mapping $(m,\lambda) \in \mathbb{R}^2 \times \mathbb{R}^3 \to j^4X_\lambda(m)$ to the sets $\Sigma^3_{S\pm}$, $\Sigma^3_{F\pm}$ and $\Sigma^3_{E\pm}$. An example of such a family in each case, called <u>standard family</u>, is given by :

$$\tilde{X}_\lambda = y\frac{\partial}{\partial x} + (\epsilon_1 x^3 + \mu_2 x + \mu_1 + y(\nu + bx + \epsilon_2 x^2))\frac{\partial}{\partial y} \tag{4}$$

where $\lambda = (\mu_1, \mu_2, \nu)$, and of course : $b>0$, $b\neq 2\sqrt{2}$, $\epsilon_{1,2} = \pm 1$.

The present article is devoted to establish several basic facts, in support of the following conjecture :

<u>Let X_λ and Y_λ be two local 3-parameter families, with X_0, Y_0 belonging to the same set : $\Sigma^3_{S\pm}$, $\Sigma^3_{F\pm}$ or $\Sigma^3_{E\pm}$. Suppose that both families are transversal to this set (in the sense defined above). Then they are fiber-C° equivalent</u> (in the sense of Chapter II).

In particular, this includes that any two standard families $\tilde{X}_\lambda$, $\tilde{Y}_\lambda$ with $\tilde{X}_0$, $\tilde{Y}_0$ in the same set $\Sigma^3_{S\pm}$, $\Sigma^3_{F\pm}$ or $\Sigma^3_{E\pm}$ will be (fiber-C°) equivalent and any generic family will be (fiber-C°) equivalent to one of the 6 standard families obtained by choosing some specific value for b : for example b=1 in the saddle and focus cases and b=3 in the elliptic case.

To proceed, we need the description of the bifurcation diagram of each generic family : how it is built and what are the different structurally stable types (open strata) in the diagram. For the proof it suffices to show that there exist just 6 types of bifurcation diagrams depending on which of the 6 sets contains X_o. Let us describe these different types.

To begin with, the bifurcation set is a topological cone with vertex at $0 \in \mathbb{R}^3$. More precisely, the parts of the bifurcation set of X_λ, adherent to $0 \in \mathbb{R}^3$, are surfaces (for the codimension 1 strata) and lines (for codimension 2 ones) which are transversal to the spheres $(\mu_1^2 + \mu_2^2 + \nu^2 = \epsilon^2)$ for ϵ small enough. So let S be such a sphere and let $\sigma \subset S$ be the intersection of the bifurcation set with S. The whole bifurcation set Σ will be a topological cone on σ, with vertex at $0 \in \mathbb{R}^3$. The codimension 1 strata of σ will consist of a finite number of lines on S ; the codimension 2 strata will be formed on one hand by end points of these lines and on the other by the intersection of 2 of these lines, finite in number. The list of all possible bifurcations of codimension 1 or 2, which appear in S, is given in Chapter III, together with their qualitative description, analytic genericity condition when necessary and also the adopted terminology.

Notice that the change in space, parameter and time $(x,y,\mu_1,\mu_2,\nu,t) \to (-x,y,-\mu_1,\mu_2,-\nu,-t)$ which transforms the equations :

$$\begin{cases} \dot{x} = y \\ \dot{y} = x^3 + \mu_2 x + \mu_1 + y\,(\nu+bx+x^2) \end{cases} \tag{5}$$

into :

$$\begin{cases} \dot{x} = y \\ \dot{y} = x^3 + \mu_2 x + \mu_1 + y\,(\nu + bx - x^2), \end{cases} \tag{6}$$

exchanges the cases Σ^3_+ and Σ^3_-. Of course the time reversal transforms stable points and cycles into unstable points and cycles. Since the description of the bifurcation diagram for (6) follows without difficulties from that for (5), it suffices to treat the cases $X_o \in \Sigma_{S+}$, Σ_{F+} and Σ_{E+}, that is $\epsilon_2 = 1$ in (4).

The bifurcation diagrams for these 3 cases are illustrated in the following 3 pictures (Figures 2, 3 and 4). The terminology for the lines and points of bifurcation is defined in Chapter IV : SN for saddle node lines, ... and so on. The subscripts l,r,i,s,sl,... refer to the position in the (x,y) plane of the bifurcation phenomenon (l : left, r : right, i : inferior, s : superior, sl : superior-left,...). To make a planar picture we deleted a point on the sphere S. This point has been chosen outside the bifurcation set, on the hemisphere $\mu_2>0$ in the saddle case and $\mu_2<0$ in the 2 other cases. In the central part of each picture, the vertical coordinate is ν; the horizontal one is μ_1, oriented to the right in the saddle case and to the left in the other two cases.
Bifurcation diagrams for the focus and saddle cases have been also proposed independently in [B,K,K], without proofs.

To complement the description of the bifurcation sets in the pictures, we give now a rough analytic definition. More details will be given, together with the proofs, in Part II of this paper. To simplify matters, we deal with the standard families (4).

The bifurcation lines fall into four different groups, according to the qualitative changes being due to bifurcations of A) singular points, B) saddle connections, C) tangencies with the boundary, or D) cycles.

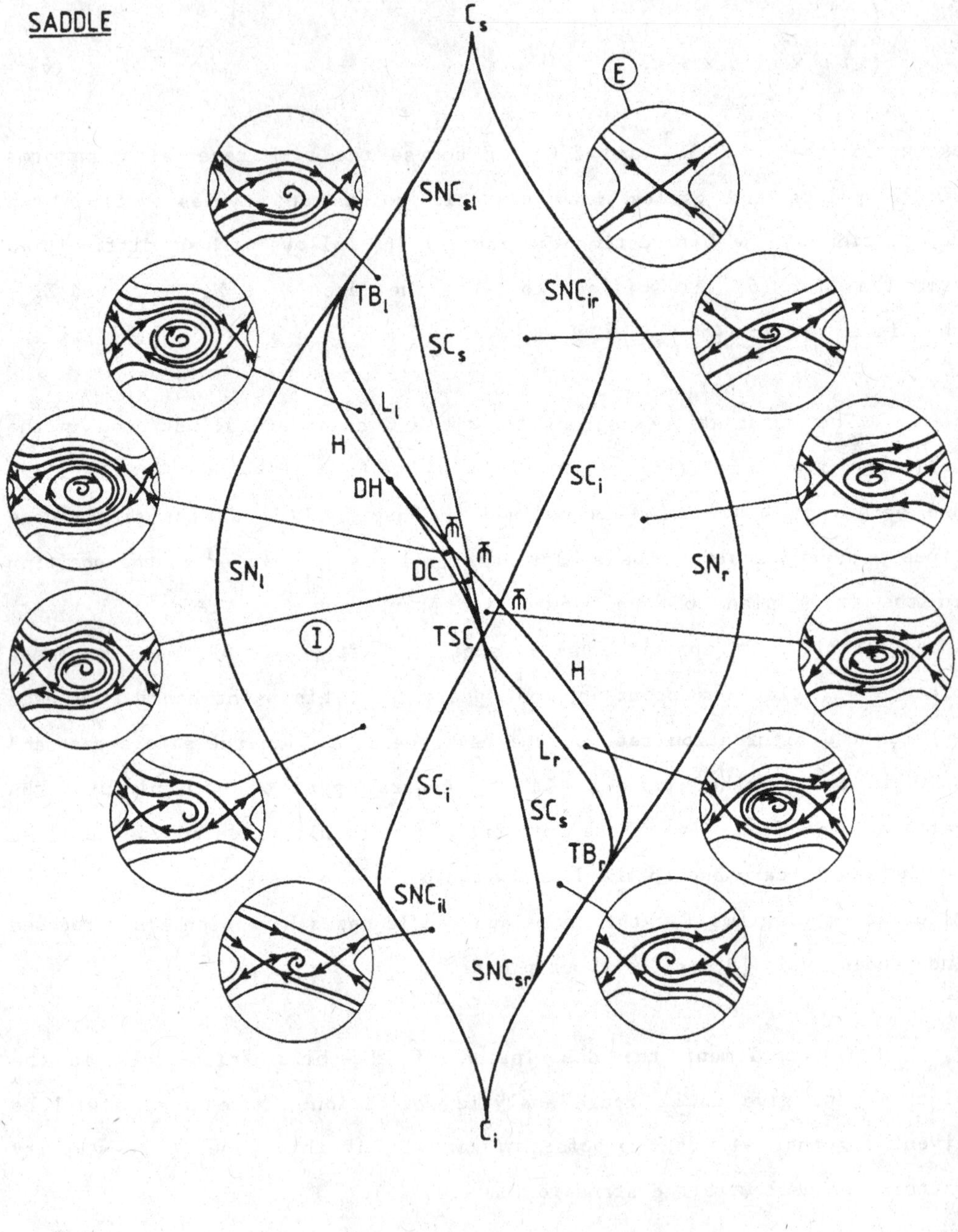
SADDLE
C_s
E
SNC_{sl}
TB_l
SNC_{ir}
SC_s
L_l
H
DH
SC_i
SN_l
DC
SN_r
I
TSC
H
L_r
SC_i
SC_s
TB_r
SNC_{il}
SNC_{sr}
C_i

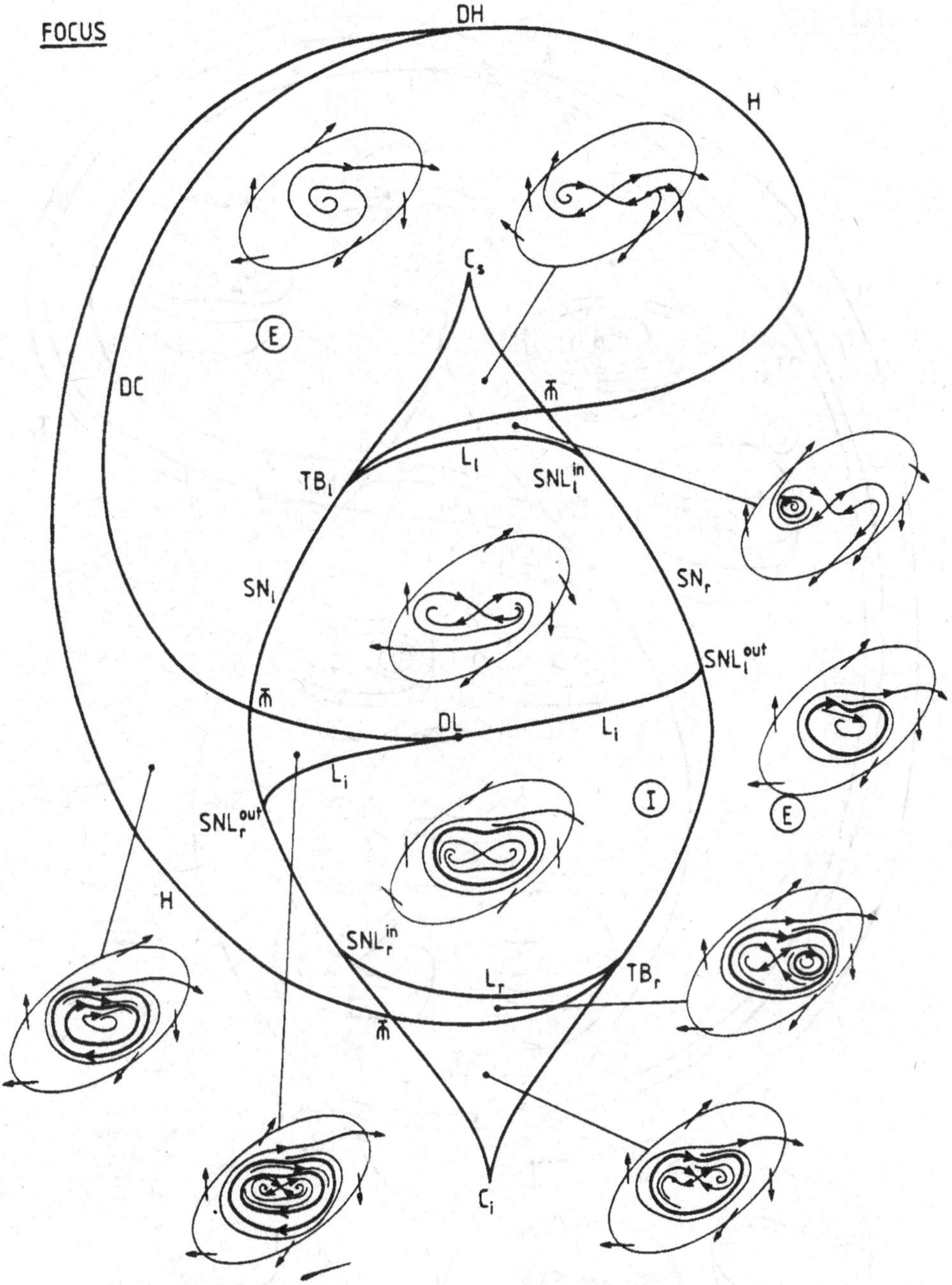
FOCUS
DH
H
C_s
E
DC
$\bar{m}$
L_l
TB_l
SNL_l^{in}
SN_l
SN_r
SNL_l^{out}
$\bar{m}$
DL
L_i
L_i
I
E
SNL_r^{out}
H
SNL_r^{in}
TB_r
L_r
$\bar{m}$
C_i

ELLIPTIC

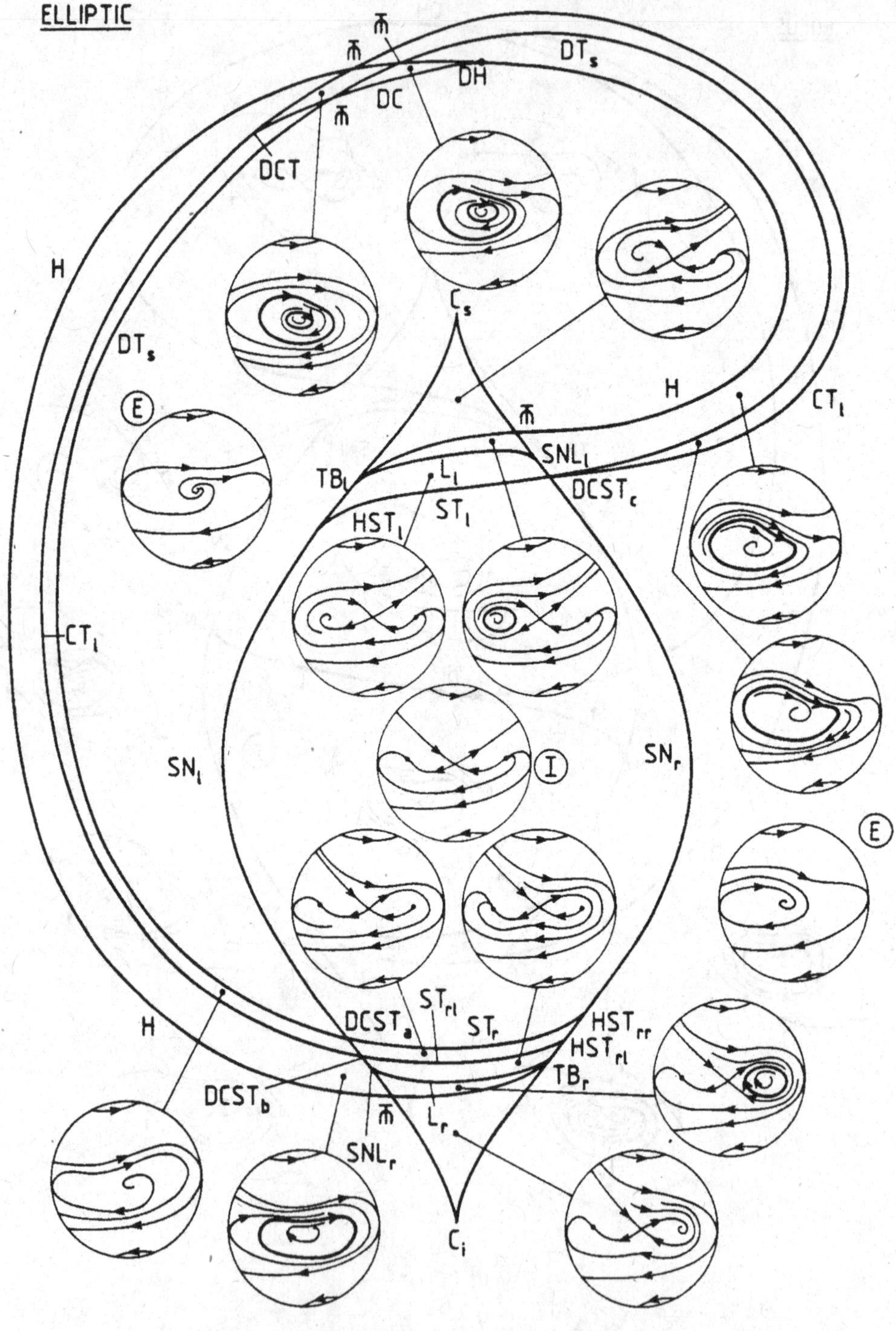

DT_s
DH
DC
DCT
H
DT_s
C_s
E
H
CT_l
SNL_l
TB_l
L_l
DCST_c
HST_l
ST_l
CT_l
I
SN_l
SN_r
E
ST_rl
DCST_a
ST_r
HST_rr
HST_rl
H
TB_r
DCST_b
SNL_r
L_r
C_i

I.2 : Codimension 1-phenomena

A. Lines corresponding to the bifurcation of singular points

The equations for singular points are : $y=0$, $\epsilon_1 x^3 + \mu_2 x + \mu_1 = 0$. The discriminant condition on the second equation : $27\mu_1^2 + 4\epsilon_1\mu_2^3 = 0$ gives 2 lines of saddle node bifurcations : $\mathbf{SN_r}$, $\mathbf{SN_1}$. These lines end in 2 cuspidal bifurcation points : $\mathbf{c_i, c_s}$. They form the corners in the big vertical lips, in each picture. In the internal region I we have 3 singular points, and in the external region E just one singular point. In the saddle case, we have a focus or a node located between 2 hyperbolic saddles for parameters in the region I. In the focus and elliptic cases, we have in region I a hyperbolic saddle located between two singularities of focus or node type.

There is also a line H of Andronov-Hopf bifurcation. It is obtained by equating to zero the trace of the family at the focus. We obtain that the trace at any singular point is zero just by elimination of x between the two equations : $\nu+bx+x^2 = 0$, $\epsilon_1 x^3 + \mu_2 x + \mu_1 = 0$. This defines a surface in the parameter space which is the image of the well known cuspidal surface $(\epsilon_1 x^3 + \mu_2 x + \mu_1 = 0)$ by the diffeomorphism : $\nu = -bx-x^2$, $\mu_1 = \mu_1$, $\mu_2 = \mu_2$. On the sphere S, this defines a tilted line h, touching respectively at points $\mathbf{TB_r}$ and $\mathbf{TB_1}$ the saddle node lines SN_r and SN_1. The restriction of the trace zero condition to foci gives H which is the part of h in region I, in the saddle case, and in region E, in the 2 other cases. The end points TB_r, TB_1 are points of Bogdanov-Takens bifurcation.

B. Lines corresponding to hyperbolic saddle separatrix connections

There are 2 lines of saddle loops $\mathbf{L_1}$, $\mathbf{L_r}$ (homoclinic connection) in each of the 3 cases. These lines are disjoint in the focus and elliptic cases, but they converge to the same point **TSC** (for "two saddles cycle") in the saddle case. They begin respectively at the Bogdanov-Takens points TB_1 and TB_r. In the focus case there exists also a line of loops $\mathbf{L_i}$ (i for inferior) connecting from below the central saddle to itself.

In the saddle case, there exist lines SC_s, SC_i of heteroclinic connections of the 2 saddle points, respectively from above (superior) and from below (inferior).

Lines of bifurcation related to boundary tangencies

These lines only appear in the elliptic case. To explain their existence we have to anticipate the definition of fiber C°-equivalence which we will state more precisely in Chapter II. We say that 2 local families X_λ and Y_λ are fiber C°-equivalent at $(0,0) \in \mathbb{R}^2 \times \mathbb{R}^3$ if there exist neighborhoods A, A_o of $0 \in \mathbb{R}^2$, B, B_o of $0 \in \mathbb{R}^3$, and a homeomorphism $\tilde{\phi} : A_o \times B_o \to A \times B$; $\tilde{\phi}(p,\lambda) = (\phi(p,\lambda), \psi(\lambda))$ such that $\forall \lambda \in B_o$, the vector field $X_\lambda | A_o$ is topologically equivalent to the field $Y_{\psi(\lambda)} \mid \phi_\lambda(A_o)$. Notice that $\phi_\lambda(.)$ is not required to be the C°-equivalence. In fact, we can suppose that $A_o \times B_o = A \times B$ and that the germs of X_λ and $Y_{\psi(\lambda)}$ are equal along ∂A_o for each $\lambda \in B_o$.

As above, B_o is supposed to be a small ball : $\partial B_o = S$. In the saddle and focus cases, we can choose the shape of the neighborhood A_o (as small as we want) such that the germ of the bifurcation set at $0 \in \mathbb{R}^3$ does not depend on A_o. In the saddle case, it suffices to choose A_o with its boundary without internal contact with X_o; in the focus case we take the boundary transversal to X_o. Take A_o to be a small disk in the normal coordinates of the Chapter III in the saddle case. In the focus case, it is easy to choose ∂A_o by approximating a segment of a spiral orbit. Of course these properties of contact with the boundary remain true for X_λ when $\|\lambda\|$ is small enough.

Now, in the elliptic case, for any choice of the neighborhood A_o there exist internal tangent points of X_o with ∂A_o. The most natural choice seems again to take for A_o a genuine small disk, in the normal coordinates (see Chapter III). So let $A_o = \{x^2 + y^2 \le \epsilon^2\}$ for a small ϵ. The vector field X_o (in the normal form of Chapter III) has just 2 internal contacts with ∂A_o at the points $a = (-\epsilon,0)$ and $b=(\epsilon,0)$ and two external contacts at the points $(0,-\epsilon)$ and $(0,\epsilon)$. Next, it is easy to prove that the positive orbit of the point b cuts ∂A_o transversally on the left of the y-axis and below the point a. (See figure 6) This follows from the assymmetry imposed by the term $x^2y \frac{\partial}{\partial y}$ and it is easily proved by comparing trajectories of X_o and of $-X_o$. We restrict B_o in such a way that these properties for ∂A_o and for the positive orbit of b remain true for each $\lambda \in B_o$. Let us illustrate the good choices of A_o in each case :

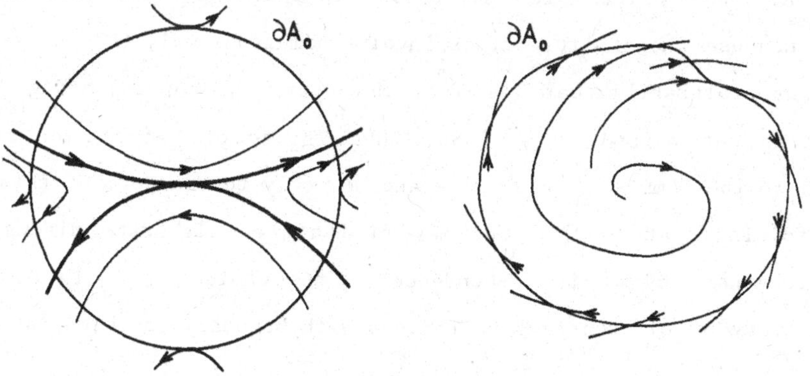

Good choices in the saddle and focus cases

Figure 5.

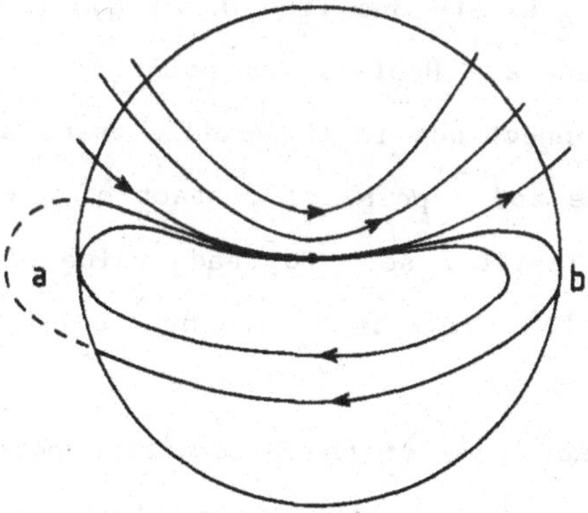

Good choice for the elliptic case

Figure 6.

The internal tangency in the elliptic case makes it inevitable to have some new codimension-1 bifurcations : the <u>separatrix tangency</u> (when a saddle separatrix passes through the points a or b), the <u>double tangency</u> (when a trajectory goes from a to b), the <u>cycle tangency</u> (when some limit cycle passes through a. A tangency of a limit cycle in b is clearly impossible. So, in the elliptic cases we have separatrix tangency lines $\mathbf{ST_1}$, $\mathbf{ST_r}$ and $\mathbf{ST_{rl}}$ (a separatrix on the right side of the unique saddle point goes to a) in the region I, a double tangency line $\mathbf{DT_s}$ and a cycle tangency line $\mathbf{CT_1}$ in the region E.

All these lines are inevitable, in the sense that they exist for any $\epsilon>0$ because these bifurcations involve simultaneously the points a or b and points arbitrarily near $0 \in \mathbb{R}^2$. Moreover, the corresponding germs of bifurcation set depend on ϵ. So, these germs of bifurcation sets are attached to the family X_λ on $A_o \times B_o$ and not only to the germ of this family (the unfoldings) at (0,0). One may find a more elaborate discussion of this phenomenon and of its dependence on the choice of A_o in Part II. A general study of bifurcations on regions with boundary has been carried out in [S_2, Te].

D. <u>Lines of double cycles</u>

In each case, there exists one line **DC** of double cycles. It is a simple arc joining a degenerate Hopf-Takens point **DH**, respectively to a point **TSC** of two saddle connections in the saddle case, a degenerate loop point **DL** in the focus case and a point of contact of a double cycle with the boundary **DCT** in the elliptic case. For each value of the parameter λ outside DC, all the periodic orbits of X_λ are hyperbolic limit cycles and their number is at most 2.

We already mentioned some of the bifurcation points (bifurcations of codimension 2). The other bifurcation points may be found on the pictures. Their designation is in accordance with their type and location (for example SNL_1^{out} designates a saddle node loop point, of outer form, located on the line SN_1). At each bifurcation point, the local situation is the generic one described in Chapter III. Apart from these points of irreducible codimension 2 type, there exist many points of transversal intersection of two codimension 1 bifurcation lines. (simply indicated by ⩚ in the pictures).

I.3 : Conic structure of the bifurcation set and rescaling

It is possible to give more precise information about the conic structure of the bifurcation set Σ in $\mathbb{R}^3$. In fact most of our knowledge about this set is obtained using one of the two following rescalings :

1. The principal rescaling. It is given by the formulas : $x = t\bar{x}$, $y = t^2\bar{y}$, $\mu_2 = t^2\bar{\mu}_2$, $\mu_1 = t^3\bar{\mu}_1$, $\nu = t\bar{\nu}$. For each $t > 0$, this defines a change of coordinates. In the new coordiates $(\bar{x},\bar{y})$ with parameters $\bar{\lambda} = (\bar{\mu}_1, \bar{\mu}_2, \bar{\nu})$ we have :

$$t^{-1}X_\lambda = \bar{y}\frac{\partial}{\partial \bar{x}} + (\epsilon_1\bar{x}^3 + \bar{\mu}_2\bar{x} + \bar{\mu}_1 + \bar{y}(\bar{\nu}+b\bar{x}))\frac{\partial}{\partial \bar{y}} + O(t) \tag{7}$$

Denote by $X^P_{\bar{\lambda}}$ the principal part of this expression. One of the effects of this rescaling is to suppress the term $yx^2\frac{\partial}{\partial y}$ in the 3-jet. Formula (7) is used in the following way : suppose that $X^P_{\bar{\lambda}}$ has a generic line or point of bifurcation B. Then, by an implicit function argument, the family X_λ has a generic surface or line of bifurcation, for t small enough, near $B \times [0,\epsilon]$ and for $((\bar{\mu}_1, \bar{\mu}_2, \bar{\nu}),t) \in \bar{S} \times [0,\epsilon]$ ($\bar{S}$ is the sphere of radius 1 in the coordinates $(\bar{\mu}_1, \bar{\mu}_2, \bar{\nu})$). For example, we will prove that all the bifurcation lines described in A, B, C above exist for $X^P_{\bar{\lambda}}$. The same is true for the points c_i, c_s, TB_1, TB_r and all the end points of the above mentioned lines (except the point TSC in the saddle case) and their transversal intersections. Let $\bar{\sigma}_1$ be the union of all these bifurcation lines and points on $\bar{S}$. For ϵ small enough, as it was already mentioned, this gives a conic subset of the bifurcation set $\Sigma_1 \subset \Sigma$ which is diffeomorphic to :

$$\{(t^3\bar{\mu}_1, t^2\bar{\mu}_2, t\bar{\nu}) \mid t \in [0,\epsilon], (\bar{\mu}_1, \bar{\mu}_2, \bar{\nu}) \in \bar{\sigma}_1\} \tag{8}$$

The principal rescaling is not sufficient to obtain the whole bifurcation set. The reason is that there exist some parameter values for which family $X^P_{\bar{\lambda}}$ is degenerate. For example in the saddle case, at the point $\bar{\lambda}_o = (-1,0,0) \in \bar{S}$ the vector field $X^P_{\bar{\lambda}_o}$ has orbits invariant under the symmetry $(\bar{x},\bar{y}) \to (-\bar{x},\bar{y})$. This is a global phenomenon of infinite codimension and no conclusion can be derived for the original family for $(\bar{\lambda},t)$ near $(\bar{\lambda}_0,0)$. Near such degenerate points in the principal rescaling the use of a secondary rescaling is most helpful.

2. The central rescaling. It is given by : $y = \tau^2 y'$, $x = \tau x'$, $\mu_2 = \tau^2\mu_2'$, $\mu_1 = \tau^4\mu_1'$, $\nu = \tau^2\nu'$. This secondary rescaling may be seen as a blowing-up of the principal one : if $\bar{\phi}$ is the principal rescaling, ϕ' the central one and ψ : $\bar{x} = x'$, $\bar{y} = y'$, $t = \tau$, $\bar{\mu}_2 = \mu_2'$, $\bar{\mu}_1 = \tau\mu_1'$, $\bar{\nu} = \tau\nu'$ denotes the blowing-up mapping, then we have that $\phi' = \bar{\phi}\circ\psi$. In the coordinates of the secondary rescaling, we have :

$$\tau^{-1}X_\lambda = y'\frac{\partial}{\partial x'} + (\epsilon_1 x'^3 + \mu_2'x' + bx'y')\frac{\partial}{\partial y'} + \tau(\mu_1' + \nu'y' + y'x'^2)\frac{\partial}{\partial y'} + o(\tau) \qquad (9)$$

Now the right hand side is a perturbation of a symmetric vector field $X^S = y'\frac{\partial}{\partial x'} + (\epsilon_1 x'^3 + \mu_2'x' + bx'y')\frac{\partial}{\partial y'}$. If $\mu_2' = \pm 1$, this vector field X^S has non degenerate singularities : saddle or center points. It follows easily that X^S is differentiably equivalent, up to a non zero multiplicative function F_b, to a Hamiltonian vector field X_{H_b} : $F_b X^S = X_{H_b}$. Such a Hamiltonian is not unique. It will be introduced and used in Part II to establish the following facts : the existence of the degenerate Hopf point DH (its location may be found directly without rescaling, but its genericity is easier to verify in the central rescaling framework), the line DC in the saddle case, the existence and genericity of the points TSC (saddle case), DL (degenerate or singular loop in the focus case) and the existence of the line DC near the points DH and DL (in the focus or elliptic cases). All these points are established in Part II using the results of H. Zoladek about Abelian integrals [Z_1,Z_2] and the methods for unfolding (singular) loops [DRS] or two-saddle cycles (see Part II and this introduction).

Let σ_2' be the union of bifurcation lines and points obtained in this way on the two planes $P_\pm' = \{\mu_2' = \pm 1\}$. Then for ϵ small enough, σ_2' defines a conic subset of the bifurcation set $\Sigma_2 \subset \Sigma$ which is diffeomorphic to :

$$\{(\tau^4 \mu_1', \tau^2\mu_1', \tau^2\nu') \mid \tau \in]0,\epsilon], (\mu_1',\nu') \in \sigma_2'\}$$

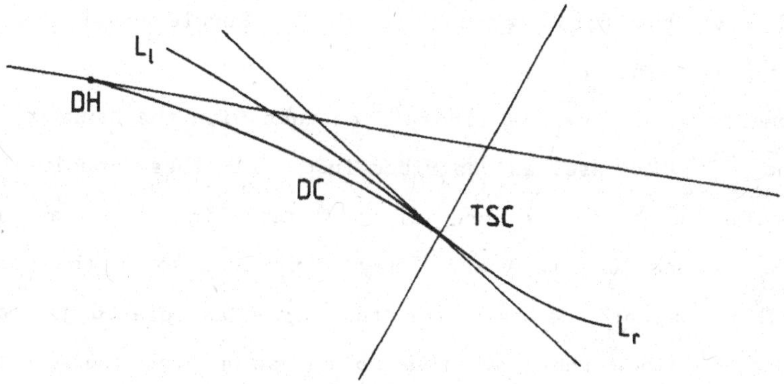

The σ_2'-set in the saddle case

Figure 7.

A word of caution : in the coordinates of the principal rescaling, the cone Σ_2 has a diameter going to zero with $t \to 0$. Of course the lines which exist in the principal rescaling appear also in the central one. However new lines and points of bifurcation (such as DH, DC in figure 7) are found, which do not exist for $X_{\bar{\lambda}}^{P}$.

I.4 : Organisation of the paper

This paper is organized as follows. The precise basic definitions and adopted notation can be found in Chapter II. In chapter III is established a useful normal form for the generic family X_λ. In Chapter IV, we give a list of the possible bifurcations of codimension 1 or 2 which are found on the sphere S, in the parameter space. In fact, as explained above, these bifurcations are obtained after rescaling. In the principal rescaling we consider generic bifurcations on the sphere $\bar{S}$. In the central rescaling, we study bifurcations obtained by perturbation of a Hamiltonian vector field. In Chapter IV are also stated the generic conditions for bifurcations in both contexts.

In Part II, we establish the existence and genericity of the lines and points as described in A (saddle-node, Andronov-Hopf, Bogdanov-Takens bifurcations). Next we study the lines described in B (saddle connections). The study is based on the rotational properties of the family relative to the parameter ν, and on the knowledge of the family around the cuspidal points c_i, c_s.
The same arguments also give the lines of contact with the boundary : DT_1, CT_1,... in the elliptic case, as described in C. All these conclusions are achieved directly without the use of any rescaling but, as already explained, the bifurcations in A and B may be found also in the principal rescaling. This is not the case for the lines associated to boundary tangencies (see C) since there we have to choose a fixed neighborhood A while the rescaled neighborhood $\bar{A}$ tends to zero with t.

Also in Part II, using the central rescaling, we perform a detailed study around TSC and so on. What remains unproved concerns essentially the existence and bifurcation of the limit cycles. This apparently difficult point is only controlled in the domain covered by the central blowing up, but it does not cover the whole region where we indicated the existence of limit cycles in pictures 2, 3, 4. In fact, we propose the simplest pictures which agree with all the known properties, established by Mathematical Analysis and also by numerical testing. Also a proof for the relative position of the Hopf line H with respect to the saddle loop lines (see B) has resisted our efforts. In Part II, Chapter VII, we also discuss the reduction of the conjecture to simpler ones. In the saddle case for instance the analysis may be reduced to verify the following :

The polynomial family

$$X^P_{\bar{\lambda}} = y\frac{\partial}{\partial x} + (x^3 - x + \bar{\mu}_1 + y(\bar{\nu} + bx))\frac{\partial}{\partial y} \qquad \bar{\lambda} = (\bar{\mu}_1, \bar{\nu}) \in \mathbb{R}^2$$

has only non-degenerate limit cycles for $\bar{\lambda} \neq (0,0)$.

This is equivalent to say that $X^P_{\bar{\lambda}}$ has a bifurcation diagram as in figure 8. It would already be interesting to prove that the lines L_l and L_r do not cut the line H outside the end points TB_l, TB_r, TSC.

In the other two cases we cannot reduce the problem to the study of X_{λ}^{P} because, as we show in VI.D.2, some limit cycles can only be studied by central rescaling, and not by the principal one. Nevertheless we expect a similar result on X_{λ}^{P} as conjectured in the saddle case.

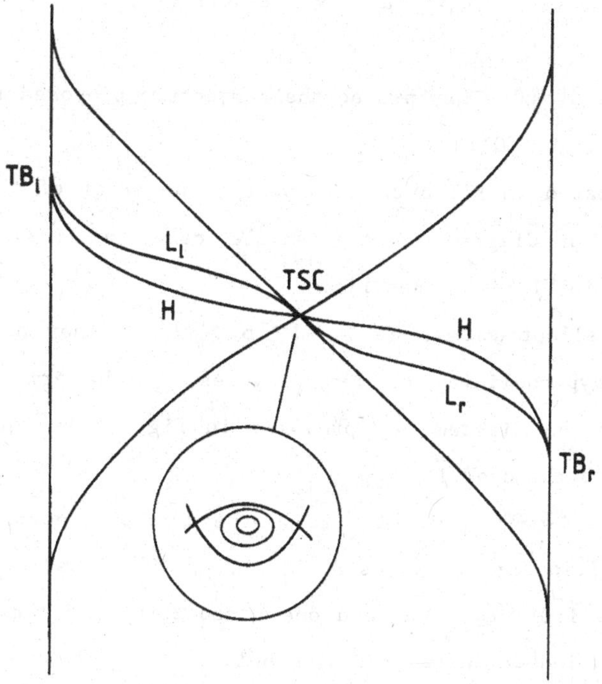

Figure 8.

Acknowledgments

We are grateful to the "Instituto de Matemática Pura e Aplicada" in Rio de Janeiro, the "Limburgs Universitair Centrum" in Diepenbeek and the "Université de Bourgogne" in Dijon for offering their hospitality, which made possible this joint work. Thanks are also due to the "Nationaal Fonds voor Wetenschappelijk Onderzoek" of Belgium and the "Conselho Nacional de Desenvolvimento Cientifico e Tecnológico" (CNPq) of Brazil for their financial support. F. Dumortier and R. Roussarie acknowledge stimulating discussions with H. Zoladek.

Recent developments

After this work was finished and available as a preprint, Christiane Rousseau and Freddy Dumortier in [D.R.] studied the families

$$X_{\bar{\lambda}}^{P} = \bar{y}\frac{\partial}{\partial \bar{x}} + (\epsilon\bar{x}^{3} + \mu_{2}\bar{x} + \mu_{1} + y(\bar{\nu} + bx))\frac{\partial}{\partial \bar{y}} ,$$

with $\epsilon = \pm 1$ and $b > 0$. They proved the conjecture proposed in I.4 for the saddle case ($\epsilon = 1$, $b > 0$).

As indicated in VII.B this completes the proof of the correctness of the bifurcation diagram in the saddle case (at least for $b \neq 1$, supposition made in VI.B.4). See fig. 2.

In the elliptic case ($\epsilon = -1$, $b > 2\sqrt{2}$), they also proved the uniqueness and hyperbolicity of limit cycles, which permits to conclude that the bifurcation diagram, as proposed in fig. 4, is correct at least inside some neighborhood of $\bar{I}$.

In the focus case ($\epsilon = -1$, $0 < b < 2\sqrt{2}$), their results permit to show that the relative position of the lines H, L_{ℓ}, L_{i} and L_{r} is as indicated in fig. 3. Moreover the limit cycle around one singularity, in the region with 3 singularities, is indeed unique and hyperbolic.

CHAPTER II : DEFINITIONS AND NOTATIONS

A k-parameter family of vector fields on $\mathbb{R}^2$, X_λ, where $\lambda \in \mathbb{R}^k$ denotes the parameter, is defined to be a vector field

$$X_\lambda = a(m,\lambda)\frac{\partial}{\partial x} + b(m,\lambda)\frac{\partial}{\partial y} \quad m = (x,y) \in \mathbb{R}^2 \tag{1}$$

where the coefficient functions a and b are C^∞ with respect to $(m,\lambda) \in \mathbb{R}^2 \times \mathbb{R}^k$.

We will study <u>local families</u> around $(0,0) \in \mathbb{R}^2 \times \mathbb{R}^k$, this means families defined on some neighborhood of (0,0), or better, <u>germs</u> of families in (0,0), since the neighborhood itself will not matter. Such a (local) family X_λ will be called a k-parameter <u>unfolding</u> (or deformation) of X_0.

Among vector fields on $\mathbb{R}^2$, we introduce the notion of <u>topological</u> (or C^0) <u>equivalence</u> : 2 vector fields X and Y are C^0-equivalent if there exists a homeomorphism h on $\mathbb{R}^2$ sending X-orbits to Y-orbits in a sense-preserving way. This notion extends to germs of vector fields in $0 \in \mathbb{R}^2$.

Related to this is the notion of <u>fiber-C°-equivalence for families of vector fields</u> : 2 families X_λ and Y_μ are called fiber-C°-equivalent if there exists a homeomorphism $\mu = \phi(\lambda)$ between the parameter spaces (of the same dimension k) and a family of homeomorphisms of $\mathbb{R}^2$ depending on the parameters λ : $h_\lambda(m)$ such that $\forall\lambda \in \mathbb{R}^k$, h_λ is a topological equivalence between X_λ and $Y_{\phi(\lambda)}$.

Notice that we do not require the equivalence to depend continuously on λ. Although we believe this to be the case in the problem here considered, it will no be included in the present study. Notice also that this relation induces an equivalence relation for local families around $(0,0) \in \mathbb{R}^2 \times \mathbb{R}^k$. It is a relation at the level of germs of families, and not of families of germs.

Suppose now that a certain family X_λ is given. The <u>bifurcation set</u> of X_λ is the smallest closed subset $\Sigma \in \mathbb{R}^k$ such that the topological type of the vector field X_λ for $\mathbb{R}^k \setminus \Sigma$ is locally constant (for the notion of C^0-equivalence).

Clearly : if 2 families are C^0-equivalent, the transformation in parameters ϕ exchanges the respective bifurcation sets.

We denote by V_o the space of germs at 0 of vector fields in $\mathbb{R}^2$ and by $J_o^N V$ the vector space of their N-jets in 0. Denote by $\pi_{PN} : J_o^P V \to J_o^N V$ (for $P \geq N \geq 0$) the natural retriction mapping and by $\pi_N : V_o \to J_o^N V$ the mapping sending a germ to its N-jet.

The natural algebraic structure of $J_o^N V$ permits us to define the notion of submanifold or (semi-) algebraic subset in $J_o^N V$; for each $\Sigma \subset J_o^N V$ we will identify Σ with its contra-images by π_{PN} and π_N in resp. $J_o^P V$ and V_o, denoting these contra-images by the same symbol Σ.

Conversely, a <u>submanifold</u> or a <u>(semi-)algebraic subset Σ</u> of <u>codimension q in V_o</u> is by definition the contra-image of a submanifold or a (semi-) algebraic subset of codimension q contained in some $J_o^N V$ and which we also denote Σ.

In the space of germs V_o, we consider the action of the group of germs of diffeomorphisms fixing 0 in $\mathbb{R}^2$ (<u>C^∞ conjugacy</u> defined by $g^* X(x) = (dg_x)^{-1} \; X(g(x))$ as well as the action of the group of pairs (f,g) consisting of the germ of a strictly positive function and the germ of a diffeomorphism fixing 0 (<u>C^∞ equivalence</u>).
This last action is defined by $((f,g).X)(x) = f(x) \; g^* X(x)$, and the group operation by $(f,g).(f',g') = (f.(f' \circ g), g' \circ g)$.

These differentiable actions on germs induce algebraic actions on each space of jets $J_o^N V$. These actions will be used in chapter III to obtain simpler expressions (normal forms).

We need the following observations : in a fixed $J_o^N V$ the subset of jets conjugate or equivalent to a certain given jet (this means an orbit of one of the corresponding group actions) form a submanifold, the set of jets conjugate or equivalent to the jets belonging to a given semi-algebraic subset form a semi-algebraic subset (theorem of Tarski-Seidenberg) [Se].

We may also define the action of C^∞ conjugacy or C^∞ equivalence on the (local) families, asking that ϕ be a (local) diffeomorphism and that $h_\lambda(m)$ be a C^∞family of C^∞diffeomorphisms (i.e. $h_\lambda(m)$ depends in a C^∞ way on m and λ). These relations will be used to obtain "normal forms" for the families X_λ.

In each point $m \in \mathbb{R}^2$ we identify the space of N-jets in m of vector fields on $\mathbb{R}^2$ to the space $J_o^N V$.
If X is a vector field on $\mathbb{R}^2$we hence obtain the N-jet mapping :

$$J^N X : \mathbb{R}^2 \to J_o^N V, \quad m \to J^N X(m)$$

If $X_\lambda(m)$ is a k-parameter family, we also consider the mapping

$$\mathbb{R}^2 \times \mathbb{R}^k \to J_o^N V, \quad (m,\lambda) \to J^N X_\lambda(m)$$

This mapping allows us to define in terms of transversality the genericity conditions on the family X_λ.

CHAPTER III : TRANSFORMATION INTO NORMAL FORM

In this chapter, we define the submanifolds $\Sigma^3_{S\pm}$, $\Sigma^3_{F\pm}$, $\Sigma^3_{E\pm} \subset J^3_0V$ and show that the 3-parameter families cutting one of these submanifolds transversally can be brought - up to C^∞ equivalence - to a simplified form - called normal form. We present this reduction to the normal form in successive steps, precribing each time for which it becomes necessary, the addition of the supplementary required hypotheses. These steps are reminiscent of those performed in [DRS] for the cusp case. This justifies the abreviated presentation that follows. Further simplification in calculations and expressions is achieved by associating to the family X_λ its dual family ω_λ of 1-forms, defined by

$$\omega_\lambda = X_\lambda \lrcorner (dx \wedge dy) \tag{1}$$

($\lrcorner$ denoting the interior product)

For $X_\lambda = a_\lambda \frac{\partial}{\partial x} + b_\lambda \frac{\partial}{\partial y}$ we have $\omega_\lambda = -b_\lambda dx + a_\lambda dy$. One can transpose to 1-forms the notion of C^∞ equivalence (conjugacy by a diffeomorphism and multiplication by a non-zero function having the same sign as the determinant of the diffeomorphism), as well as the corresponding notion at the germ level.

Two families are C^∞ equivalent if and only if the dual families of 1-forms are C^∞ equivalent.

Let us start with a k-parameter family X_λ with the unique hypothesis :

(Hyp 1) $J^1X_0(0)$ is linearly conjugate to $y \frac{\partial}{\partial x}$.

So, up to linear conjugacy, we may suppose : $J^1X_0(0) = y \frac{\partial}{\partial x}$. This condition defines an algebraic submanifold of codimension 4 in J^1_0V and, as we know from $[A_2]$, $[B_1]$, $[T_2]$, the family X_λ can be put in the following normal form by C^∞ equivalence (even C^∞ conjugacy) :

$$X_\lambda \sim y \frac{\partial}{\partial x} + (F(x,\lambda) + yG(x,\lambda)) \frac{\partial}{\partial y} + Q_1 \frac{\partial}{\partial x} + Q_2 \frac{\partial}{\partial y} , \tag{2}$$

where $\sim$ is C^∞ equivalence, Q_1 and Q_2 are of order $O((\| m \| + \| \lambda \|)^N)$ for a certain N that one can choose arbitrarily big, $m = (x,y)$, $\| \cdot \|$ are any norms on $\mathbb{R}^2$ and $\mathbb{R}^k$, F and G are C^∞ functions in (x,λ) and we may suppose that they are polynomial of degree N in x. The equations of the orbits of (2) are :

$$\begin{cases} \dot{x} = y + Q_1 \\ \dot{y} = F(x,\lambda) + yG(x,\lambda) + Q_2 \end{cases} \tag{2'}$$

<u>1st step</u> : <u>Reduction to a differential equation of 2nd order</u>

The following λ-dependent coordinate change :

$Y = y + Q_1$, $X = x$ transforms equations (2') into :

$$\begin{cases} \dot{X} = Y \\ \dot{Y} = F(X,\lambda) + YG(X,\lambda) + Q_2'(X,Y,\lambda) \end{cases} \tag{3}$$

Where $Q_2' = O((\| M \| + \| \lambda \|)^{N-1})$, $M = (X,Y)$

Changing N-1 into N and (X,Y) into (x,y), we find back the expression (2') with $Q_1 = 0$. So, using a C^∞ equivalence, we have changed the original family of differential equations into a parameter - dependent differential equation of 2nd order :

$$\ddot{x} = F(x,\lambda) + \dot{x}G(x,\lambda) + Q(x,\dot{x},\lambda), \tag{4}$$

where Q is of order N, $F(0,0) = \frac{\partial F}{\partial x}(0,0) = G(0,0) = 0$

<u>2nd step</u> : <u>Division of the term Q by y^2</u>

We can develop Q in powers of y :

$$Q(x,y,\lambda) = \tilde{F}(x,\lambda) + y\tilde{G}(x,\lambda) + y^2\tilde{Q}(x,y,\lambda)$$

So, with an evident change of notation, we obtain that (4) is C^∞ equivalent to :

$$\ddot{x} = F(x,\lambda) + \dot{x}G(x,\lambda) + (\dot{x})^2 Q(x,\dot{x},\lambda) \tag{5}$$

where Q if of order N, $F(0,0) = \frac{\partial F}{\partial x}(0,0) = G(0,0) = 0$

We now introduce our second hypothesis :

(Hyp 2) $\frac{\partial^2 F}{\partial x^2}(0,0) = 0$, $\frac{\partial^3 F}{\partial x^3}(0,0) \neq 0$ and $\frac{\partial G}{\partial x}(0,0) \neq 0$.

<u>3rd step</u> : <u>Reduction of $F(x,\lambda)$ to $\epsilon_1 x^3 + \mu_2(\lambda)x + \mu_1(\lambda)$ ($\epsilon_1 = \pm 1$)</u>

Hypothesis 2 implies that $F(x,0)dx$ is the differential of a function of order 4 at x=0. Such a function admits as universal unfolding :

$$\epsilon_1 \frac{x^4}{4} + \mu_2 \frac{x^2}{2} + \mu_1 x ,$$

where the term $\epsilon_1 = \pm 1$ has the sign of $\frac{\partial^3 F}{\partial x^3}(0,0)$.

Hence, there exists a differentiable mapping $\mu(\lambda) = (\mu_1(\lambda), \mu_2(\lambda))$ and a family of diffeomorphisms depending on the parameter λ :

$U_\lambda(x) = u(\lambda)x + O(x^2) + O(\|\lambda\|)$ such that

$U_\lambda^*(F(x,\lambda)dx) = (\epsilon_1 x^3 + \mu_2(\lambda)x + \mu_1(\lambda))dx$, with $\mu_1(0) = \mu_2(0) = 0$.

Performing the C^∞ equivalence $\tilde{U}_\lambda : (x,y) \to (U_\lambda(x),y)$ on the dual family ω_λ, we get :

$$\omega_\lambda \sim y\,dy - [(\epsilon_1 x^3 + \mu_2(\lambda)x + \mu_1(\lambda) + y\,\tilde{G}(x,\lambda) + y^2\,\tilde{Q}(x,y,\lambda)]\,dx \tag{6}$$

As $\tilde{G}dx = U_\lambda^*(Gdx)$, $\tilde{Q}dx = U_\lambda^*(Qdx)$, $U(x,0) = u(0)x + O(x^2)$ and $U(x,\lambda) = O((|x| + \|\lambda\|))$, the functions $\tilde{G}$, $\tilde{Q}$ have the same properties as above : $\frac{\partial \tilde{G}}{\partial x}(0,0) \neq 0$ and $\tilde{Q} = O((\|m\| + \|\lambda\|)^N)$.

So, with an obvious change of notation we have :

$$X_\lambda \sim y\frac{\partial}{\partial x} + (\epsilon_1 x^3 + \mu_2(\lambda)x + \mu_1(\lambda) + yG(x,\lambda) + y^2 Q(x,y,\lambda))\frac{\partial}{\partial y}, \qquad (7)$$

with $\frac{\partial G}{\partial x}(0,0) \neq 0$ and Q of order N.

Consider now the 4-jet of X_o :

$$j^4 X_o(0) = y\frac{\partial}{\partial x} + (\epsilon_1 x^3 + bxy + cx^2 y + c'x^3 y)\frac{\partial}{\partial y} \qquad (8)$$

with $\epsilon_1 = \pm 1$, $b = \frac{\partial G}{\partial x}(0,0)$ and $c = \frac{1}{2}\frac{\partial^2 G}{\partial x^2}(0,0)$.
As it was recalled in the introduction, the topological type of X_o is determined by its 3-jet if $b \neq 0$ and $b \neq 2\sqrt{2}$ in case $\epsilon_1 = -1$. We moreover impose the condition $c \neq 0$.

It is an easy calculation to show that from expression (3) follows that $c = a - \frac{3bd}{5\epsilon_1}$, so that $c \neq 0$ is the same as condition (3').
These two additional conditions constitute our third hypothesis :

(Hyp 3) $\frac{\partial^2 G}{\partial x^2}(0,0) \neq 0$, $\frac{\partial G}{\partial x}(0,0) \neq 0$ and $\frac{\partial G}{\partial x}(0,0) \neq 2\sqrt{2}$ (in case $\epsilon_1 = -1$)

<u>4th step</u> : <u>Reduction to $G(x,\lambda) = \nu(\lambda) + b(\lambda)x + \epsilon_2 x^2 + O(x^3)$ with $b(0) = b > 0$ and $\epsilon_2 = \pm 1$.</u>

Let $G(x,\lambda) = \nu(\lambda) + b(\lambda)x + c(\lambda)x^2 + O(x^3)$.
We have that $\nu(0) = 0$, $b(0) \neq 0$ and $c(0) \neq 0$.
Consider the linear coordinate change, depending on the parameter λ :
$U_\lambda : (x,y) \to (\alpha(\lambda)x, \beta(\lambda)y)$. Applying it to X_λ, we obtain :

$$(U_\lambda)_* (X_\lambda) = \frac{\beta}{\alpha} y \frac{\partial}{\partial x} + \frac{1}{\beta} (\epsilon_1 \alpha^3 x^3 + \mu_2 \alpha x + \mu_1 + \beta y (\nu + \alpha b x + \alpha^2 c x^2 + O(x^3))$$

$$+ \beta^2 y^2 Q(\alpha x, \beta y, \lambda)) \frac{\partial}{\partial y}$$

Taking $|\beta| = \alpha^2$, Sign (α) = Sign (β) = Sign $b(0)$ and $|\alpha(\lambda)| = \frac{1}{\sqrt{c(\lambda)}}$, we see that U_λ is a C^∞ equivalence which transforms the family X_λ into a new one of the same form (7), but which enjoys the desired properties on $G(x,\lambda)$. Observe that we have only possibly changed the sign of $b = \frac{\partial G}{\partial x}(0,0)$, which is coherent with the assumption $\frac{\partial G}{\partial x}(0,0) \neq \pm 2\sqrt{2}$.

Let us recall that the submanifolds $\Sigma_{S\pm}$, $\Sigma_{F\pm}$, $\Sigma_{E\pm}$ of V_o are defined by :

$$\Sigma_{S\pm} = \{X \in V_o \mid j^4 X(0) \sim y \frac{\partial}{\partial x} + (x^3 + bxy \pm yx^2 + fx^3 y) \frac{\partial}{\partial y}, \ b>0, \ f \in \mathbb{R}\}$$

$$\Sigma_{F\pm} = \{X \in V_o \mid j^4 X(0) \sim y \frac{\partial}{\partial x} + (-x^3 + bxy \pm yx^2 + fx^3 y) \frac{\partial}{\partial y}, \ 0<b<2\sqrt{2}, \ f \in \mathbb{R}\}$$

$$\Sigma_{E\pm} = \{X \in V_o \mid j^4 X(0) \sim y \frac{\partial}{\partial x} + (-x^3 + bxy \pm yx^2 + fx^3 y) \frac{\partial}{\partial y}, \ b>2\sqrt{2}, \ f \in \mathbb{R}\}$$

The hypotheses 1, 2, 3 made so far, imply that X_0 belongs to the union Σ^3_{SFE} of these 6 submanifolds of V_o.
The last hypothesis concerns the genericity of the family X_λ.

<u>5th step</u> : <u>Genericity of the family</u> X_λ

Suppose now that $\lambda \in \mathbb{R}^3$: $\lambda = (\lambda_1, \lambda_2, \lambda_3)$. The transversality condition of $j^4 X_\lambda(m)$ with respect to Σ^3_{SFE} amounts to :

$$\text{(Hyp 4)} \quad \frac{D(\mu_1, \mu_2, \nu)}{D(\lambda_1, \lambda_2, \lambda_3)}(0) \neq 0$$

Under this generic condition, we may choose $\lambda = (\mu_1, \mu_2, \nu)$, after a diffeomorphic change in the parameter space. We finally obtain the definitive normal form, obtained using C^∞ equivalence :

$$X_\lambda = y\frac{\partial}{\partial x} + (\epsilon_1 x^3 + \mu_2 x + \mu_1 + y(\nu + b(\lambda)x + \epsilon_2 x^2 + x^3 h(x,\lambda)) + y^2 Q(x,y,\lambda))\frac{\partial}{\partial y} \tag{9}$$

where $\epsilon_1, \epsilon_2 = \pm 1$, $\lambda = (\mu_1, \mu_2, \nu)$ is the parameter, $b(\lambda)$ $(b(0) > 0)$ is a C^∞ function and Q is a C^∞ function, of order N in (x,y,λ) where N is arbitrarily high.

6th step : As already observed in the introduction one can change $\epsilon_2 = -1$ into $\epsilon_2 = 1$ by means of the coordinate change $(x,y,\mu_1,\mu_2,\nu,t) \to (-x,y,-\mu_1,\mu_2,-\nu,-t)$, which does not modify the other essential features : ϵ_1 keeps it sign, $b>0$.

This change is no longer a C^∞-equivalence since we admit a reversal of time.

CHAPTER IV : BIFURCATIONS OF CODIMENSION 1 AND 2

IV.1 : Generalities

In this chapter we establish the list of all bifurcations of codimensions 1 and 2 used in this study. Most of them are well known [A, A.L., S, Sc, Te, G.H., H.C.]. Below follows a quick review of the aspects most relevant for our work, emphasizing the less known cases.
The bifurcations we encounter may be local (i.e. : Andronov-Hopf bifurcation) or global (i.e. : connection between saddle points). Next, we obtain our results after rescaling. After this process of rescaling, we have a family $X_{\bar{\lambda},t}$ where t is a small parameter and $\bar{\lambda} \in K$, some compact subset of S^2 or $\mathbb{R}^2$. Depending on the nature of the $\bar{\lambda}$-family $X_{\bar{\lambda},0}$ around some value $\bar{\lambda}_o$, the study splits into two cases :

- The generic case : Let $\bar{\lambda}_o$ be a generic bifurcation value for the family $X_{\bar{\lambda},0}$. Then, the bifurcation set for $X_{\bar{\lambda},0}$ is given by transversality conditions and, using an implicit function argument we obtain for $X_{\bar{\lambda},t}$ and small t, a bifurcation set with the same codimension. If σ is the local bifurcation set of $X_{\bar{\lambda},0}$ at $\bar{\lambda}_o$, then the local bifurcation set for $X_{\bar{\lambda},t}$ is diffeomorphic to $\sigma \times [0,\epsilon]$ for $t \in [0,\epsilon]$, ϵ small enough.

- The perturbed Hamiltonian (P.H.) case :
  Here, up to multiplication of the family by a positive C^∞ function, we have that $X_{\bar{\lambda},0}$ is a Hamiltonian vector field, independent of $\bar{\lambda}$. Let $\omega_{\bar{\lambda},t}$ be the dual form of $X_{\bar{\lambda},t}$ (see Chapter III). Then $\omega_{\bar{\lambda},0} = dH$ for some C^∞ function H, and we can expand :

$$\omega_{\bar{\lambda},t} = dH - t\omega_D(\bar{\lambda}) + o(t) \qquad (1)$$

To simplify the discussion, suppose that for $\forall t$, $\omega_{\bar{\lambda},t}$ has the same singular points as dH. Let $\sigma = [a,b]$ be a segment in the phase plane which at each point of $]a,b]$ is transversal to a closed component of a level curve of H. Suppose that at a it is also transversal to a closed cycle or it is a non degenerate center of H. We parametrize σ by means of the value of H : $\sigma \simeq [\alpha,\beta] = [H(a), H(b)]$. Most of the results we use in the P-H-case are derived from the following well known lemma [A.L.] :

Lemma (Perturbation lemma)

Let $\omega_{\bar{\lambda},t}$, H, σ be as above. Let K be a compact subset in the parameter space of $\bar{\lambda}$. Then there exists a T(K) > 0 such that for all $(\bar{\lambda},t) \in K \times [0,T(K)]$:

1) The vector field $X_{\bar{\lambda},t}$ is transversal to $]\alpha,\beta]$
2) The Poincaré map $P_{\bar{\lambda},t}(h)$ of $X_{\bar{\lambda},t}$, or its inverse $P^{-1}_{\bar{\lambda},t}(h)$ is defined on $[\alpha,\beta]$.
3) For $h \in [\alpha,\beta]$, the coordinate defined by the value of H, it holds that

$$P_{\bar{\lambda},t}(h) = h + t\int_{\gamma_h} \omega_D(\bar{\lambda}) + o(t) \tag{2}$$

where γ_h is the compact component of $\{H=h\}$ passing through the point $h \in [\alpha,\beta]$, clockwise oriented for the integration.

Let $I(h,\bar{\lambda}) = \int_{\gamma_h} \omega_D(\bar{\lambda})$ be the Abelian integral giving the first order term in formula (2). The fixed points of $P_{\bar{\lambda},t}$ are the zeroes of the function :

$$G(h,\bar{\lambda},t) = \frac{P_{t,\bar{\lambda}}(h) - h}{t} = I(h,\bar{\lambda}) + O(t) \tag{3}$$

Here, O(t) is a C^∞ function in $(h,\bar{\lambda},t)$ of order t.

Using formula (3) it is easy to find conditions for existence and genericity of Hopf bifurcations (at $h=\alpha$) and of limit cycle bifurcations (at $h \neq \alpha$) in any codimension, at least for small values of $t \neq 0$.

Suppose now that the endpoint β belongs to Γ, where Γ is some hyperbolic singular cycle of H. That is a connected compact piece of $\{H=\beta\}$ made of saddle points and regular arcs. In this paper we will encounter the saddle loop (singular cycle with just 1 saddle) and the two-saddles cycle (cycle with 2 saddles connected by 2 arcs). Then, the formulas (2) and (3) are not sufficient to study the bifurcations of $X_{\bar{\lambda},t}$ near Γ, because the mapping $P_{\bar{\lambda},t}$ is no longer differentiable at $h = \beta$ for $t = 0$. A direct study of $P_{\bar{\lambda},t}$ is needed to relate the bifurcations of $X_{\bar{\lambda},t}$ to an expansion of the Abelian integral near $\beta = 0$.

In what follows we do not need to specify the dimension of the parameter space for λ (in the generic case) or for $\bar{\lambda}$ (in the P-H case). Each bifurcation corresponds to some stratum Σ in the space of vector fields. We give its local equations (conditions expressing that $X_{\lambda_0} \in \Sigma$). These local equations reduce to one equation $\mu(\lambda) = 0$ for the codimension 1 bifurcations or to 2 equations $\mu(\lambda) = \nu(\lambda) = 0$ for codimension 2 bifurcations. In each case we suppose that the family is generic, i.e. transversal to the stratum Σ. This is equivalent to say that the defining equations are independent; i.e. $d\mu(\lambda_0) \neq 0$, for codimension 1, and that $d\mu(\lambda_0)$, $d\nu(\lambda_0)$ form a set of 2 independent 1-forms, for codimension 2. We will not recall these genericity conditions explicity in each case.
A consequence is that we can suppose that μ or (μ,ν) are some coordinates for λ and that the local bifurcation diagram is diffeomorphic to the product of the bifurcation set in the μ or (μ,ν) coordinate-space with the space of the other parameters. So it will suffice to describe the 1-parameter family (given by the variation of μ) or the 2-parameter family (given by the variation of μ,ν) which may be considered as the versal unfolding of X_{λ_0}.

In the P-H-case we will work around some value $(\bar{\lambda}_0,0)$. The equations are given in the $\bar{\lambda}$-parameter and, as above, the genericity condition reduces to the independence of these equations. If $\bar{\sigma}$ is the local submanifold which they define in the $\bar{\lambda}$-space, then the bifurcation set itself is diffeomorphic to $\bar{\sigma} \times]0,\epsilon]$ for ϵ small enough, by a diffeomorphism $(\bar{\lambda},t) \to (h_t(\lambda),t)$ where h_t tends to the identity when $t \to 0$. The bifurcations are described up to some C^∞ equivalence, which we do not make explicit.

IV.2 : Codimension 1 bifurcations

1. Andronov-Hopf bifurcation : (H)

We may suppose : $j^1 X_{\lambda_o}(0) = \pm (y \frac{\partial}{\partial x} - x \frac{\partial}{\partial y})$ ($0 \in \mathbb{R}^2$ is a non-degenerate center for the linear part of X_{λ_o}). Let $c(\lambda)$ be the unique singularity of X_λ near 0, and $\mu(\lambda) \pm i\nu(\lambda)$ its eigenvalues ($\mu(\lambda_o) = 0$). Suppose also that $\pm j^\infty X_{\lambda_o}(0) = y \frac{\partial}{\partial x} - (x + \Sigma_{i,j}\, b_{ij}\, x^i y^j) \frac{\partial}{\partial y}$.

Then, we have the following characterization :

Generic case :

The bifurcation set is given $\mu(\lambda) = 0$ and $L_1 \neq 0$, where L_1 is the first Lyapounov exponent; $L_1 \neq 0$ when $-3b_{03} - b_{21} - b_{11}(b_{02} + b_{20}) \neq 0$ (see [A.L.] and also [G.H.]).

P-H-case :

Suppose that H=0 at a center point. Let us write the h-expansion of the integral I at 0 : $I(h,\bar{\lambda}) = \bar{\mu}_1(\bar{\lambda})h + \bar{\mu}_2(\bar{\lambda})h^2 + O(h^3)$. The bifurcation set is given by $\bar{\mu}_1(\bar{\lambda}) = 0$ and $\bar{\mu}_2(\bar{\lambda}_o) \neq 0$.

2. Saddle-node (SN) and Saddle-node loop (SNL_1).

We may suppose that $j^1 X_{\lambda_o}(0) = \pm y \frac{\partial}{\partial y}$ and that Ox is a center manifold of X_{λ_o}, invariant for each λ. Then we may suppose that

$X_\lambda = \pm y(1 + O(|m|)) \frac{\partial}{\partial y} + (f(x,\lambda) + O(y)) \frac{\partial}{\partial x}$ with :

$$f(0,\lambda_o) = \frac{\partial f}{\partial x}(0,\lambda_o) = 0 \quad \text{and} \quad \frac{\partial^2 f}{\partial x^2}(0,\lambda_o) \neq 0$$

The bifurcation set is given by : $f(x,\lambda) = \frac{\partial f}{\partial x}(x,\lambda) = 0$.

Remark

Along certain SN-lines the bifurcation has no other effect on the ω-limit set than the disappearance of two singularities.
Along other lines the coalescence and disappearance of the two singularities gives rise to a unique hyperbolic limit cycle (attracting or expanding). This happens when at the SN-point the isolated center manifold returns to the singularity, through the interior of the nodal sector, forming a saddle-node loop of codimension 1 (SNL_1).

3. Saddle loop (L)

We suppose that X_{λ_o} has a hyperbolic saddle $s(\lambda_o)$ with a homoclinic connection Γ. Let σ be a segment transverse to Γ. Let $s(\lambda)$ be the unique singularity of X_λ near $s(\lambda_o)$ and $-u(\lambda)$, $v(\lambda)$ its eigenvalues ($u(\lambda), v(\lambda)>0$). Let $W^s(\lambda)$, $W^u(\lambda)$ be the stable and unstable separatrices of X_λ, near Γ ($\Gamma = W^s(\lambda_o) = W^u(\lambda_o)$).
Let $\{a(\lambda)\} = W^s(\lambda) \cap \sigma$, $\{b(\lambda)\} = W^u(\lambda) \cap \sigma$ and $\mu(\lambda) = a(\lambda) - b(\lambda)$ ($\mu(\lambda_o) = 0$). See Fig. 9.

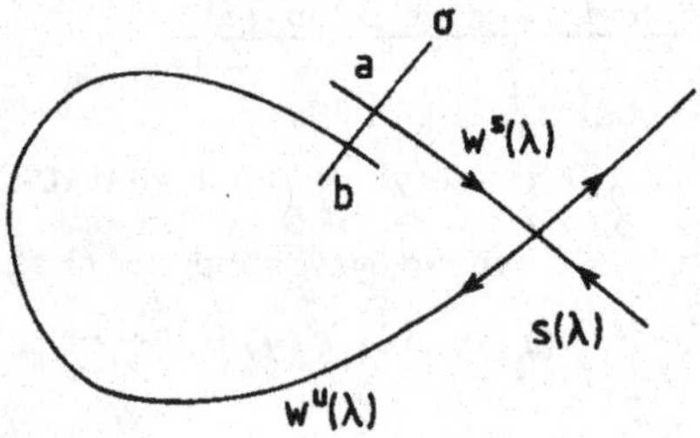

Figure 9.

Generic case :

The bifurcation set is given by $\mu(\lambda) = 0$, $\dfrac{u(\lambda_o)}{v(\lambda_o)} \neq 1$

P-H-case :

We suppose that the $\{H = \beta\}$ contains a loop Γ with a hyperbolic saddle s. Then $X_{\bar{\lambda},t}$ also has a saddle at s (see the introduction of the chapter). Let $-u(t,\bar{\lambda})$, $v(t,\bar{\lambda})$ be the eigenvalues of this saddle. We have

$$\frac{u(t,\bar{\lambda})}{v(t,\lambda)} = 1 - t\alpha(\bar{\lambda}) + o(t).$$

Let also $I_\beta(\bar{\lambda}) = I(\beta,\bar{\lambda}) = \int_\Gamma w_D(\bar{\lambda})$ (integral of the loop).
Then, the bifurcation set is given by : $I_\beta(\bar{\lambda}) = 0$, with $\alpha(\bar{\lambda}_o) \neq 0$.

4. Saddle connection (SC)

Here, X_{λ_o} has 2 hyperbolic saddles $s_1(\lambda_o)$, $s_2(\lambda_o)$, connected by a separatrix Γ which coincides with an unstable manifold $W^u(\lambda_o)$ of $s_1(\lambda_o)$ and a stable one $W^s(\lambda_o)$ of $s_2(\lambda_o)$. Let σ be a transversal to Γ, let $s_1(\lambda)$, $s_2(\lambda)$ be the unique singularities of X_λ near $s_1(\lambda_o)$, $s_2(\lambda_o)$, $W^s(\lambda)$ and $W^u(\lambda)$ the invariant manifolds of $s_1(\lambda)$, $s_2(\lambda)$ near Γ.
Let $\{a(\lambda)\} = \sigma \cap W^s(\lambda)$, $\{b(\lambda)\} = \sigma \cap W^u(\lambda)$ and $\mu(\lambda) = a(\lambda) - b(\lambda)$ with $\{a(\lambda_o)\} = \{b(\lambda_o)\} = \sigma \cap \Gamma$. See Fig. 10.

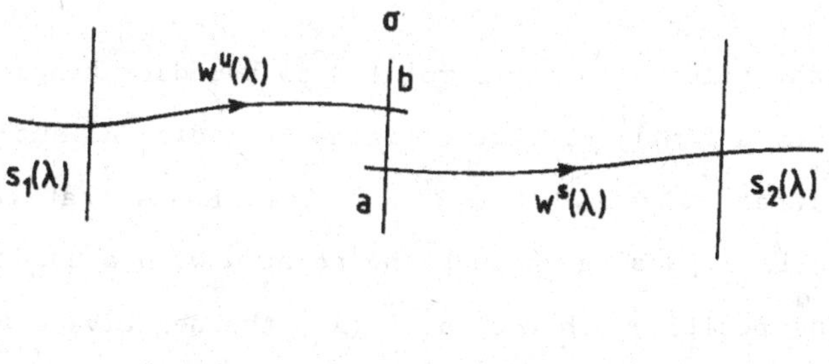

Figure 10.

Generic case :

The bifurcation set is given by $\mu(\lambda) = 0$.

P-H-case :

We suppose that the function H has a connection Γ between two saddle points s_1, s_2 contained in the level $\{H = \beta\}$. Let $I(\bar{\lambda}) = \int_\Gamma \omega_D(\bar{\lambda})$ (Notice that the convergence follows from $\omega_D(\bar{\lambda})(s_1) = \omega_D(\bar{\lambda})(s_2) = 0$.

Then, the equation of the bifurcation set is $I(\bar{\lambda}) = 0$.

5. Double cycle (DC)

We suppose that X_{λ_o} has a semi-stable limit cycle Γ. Let σ be a transversal to Γ, and $P(h,\lambda)$ the return map on σ defined for (h,λ) near (h_o,λ_o) where $\{h_o\} = \sigma \cap \Gamma$.

Generic case :

The equations for the bifurcation set are : $P(h,\lambda) = \frac{\partial P}{\partial h}(h,\lambda) = 0$ and $\frac{\partial^2 P}{\partial h^2}(h_o,\lambda_o) \neq 0$.

P-H-case :

Let $I(h,\bar{\lambda})$ be the Abelian integral associated to the family. We suppose that for some $h_o \neq \alpha$: $I(h_o,\bar{\lambda}_o) = \frac{\partial I}{\partial h}(h_o,\bar{\lambda}_o) = 0$ and that $\frac{\partial^2 I}{\partial h^2}(h_o,\bar{\lambda}_o) \neq 0$. Then the equation for the bifurcation set is $I = \frac{\partial I}{\partial h} = 0$.

The 3 last bifurcations are related to boundary tangencies. So, we suppose that X_λ is a family of vector fields on a disk A, such that X_{λ_o} has some quadratic inner tangency at $\alpha \in A$. This means that there exists a trajectory of X_{λ_o} in A passing through the point α with a quadratic contact. Let $\Gamma_+(\lambda_o)$ be the positive orbit of α, $\Gamma_-(\lambda_o)$ the negative orbit of α. For all λ near λ_o, the vector field X_λ has a unique tangency $\alpha(\lambda)$ near $\alpha = \alpha(\lambda_o)$. This tangency is again a quadratic inner one. Let $\Gamma_+(\lambda)$, $\Gamma_-(\lambda)$ be the corresponding half orbits. See fig. 11.

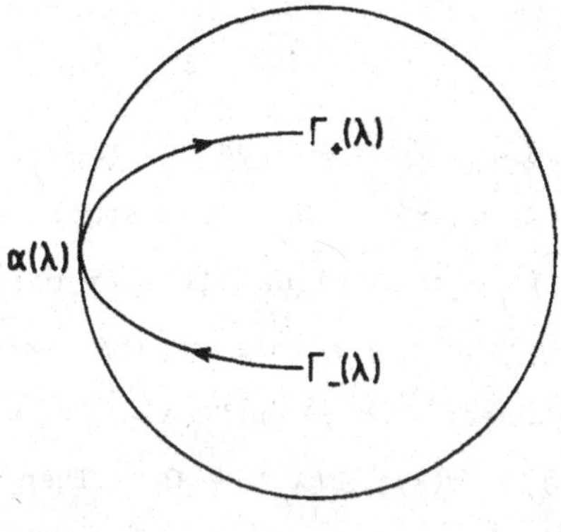

Figure 11.

6. Cycle Tangency (CT)

X_{λ_o} has a limit cycle Γ passing through α. So $\Gamma_+(\lambda_o) = \Gamma_-(\lambda_o) = \Gamma$. Let σ be a transversal to Γ. Then $\Gamma_+(\lambda)$ cuts σ at a point $a(\lambda)$ near $\Gamma \cap \sigma$ and $\Gamma_-(\lambda)$ cuts σ at a point $b(\lambda)$ near $\Gamma \cap \sigma$.

Let $\mu(\lambda) = a(\lambda) - b(\lambda)$; $\mu(\lambda_o) = 0$.

Then the bifurcation set is given by $\mu(\lambda) = 0$ and the condition that Γ is normally hyperbolic. For instance, let us suppose that Γ is expansive and that there exists a parameter in λ, λ_1 such that $\frac{\partial\mu}{\partial\lambda_1}(\lambda_o) > 0$. Then, the bifurcation in function of λ_1 is given in Fig. 12, where λ_1 is increasing from left to the right.

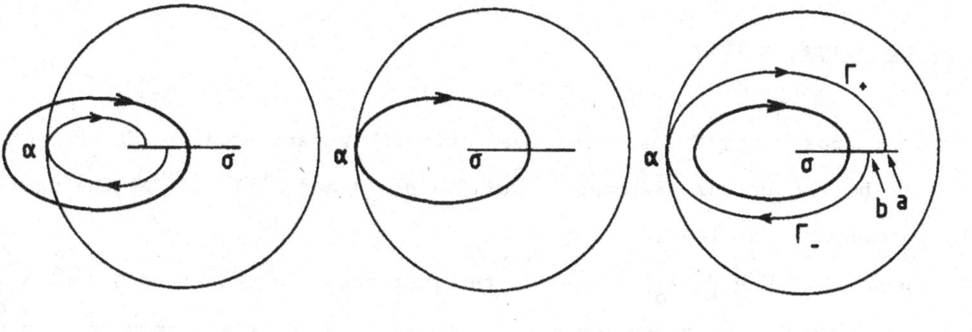

Figure 12.

7. Separatrix tangency (ST)

X_{λ_o} has a saddle point $s(\lambda_o)$ with a separatrix passing through α. For example $\Gamma = \Gamma_+(\lambda_o) = W^s(\lambda_o)$ where $W^s(\lambda_o)$ is a stable separatrix. Let σ be transverse to Γ. Let $s(\lambda)$ be the unique singular point of X_λ near $s(\lambda_o)$. It is a saddle. Let $W^s(\lambda)$ be the stable separatrix of $s(\lambda)$ near $W^s(\lambda_o)$, $\{b(\lambda)\} = \sigma \cap W^s(\lambda)$ (point near $\Gamma \cap \sigma$) and $\{a(\lambda)\} = \sigma \cap \Gamma^+(\lambda)$ (point near $\Gamma \cap \sigma$). Let $\mu(\lambda) = a(\lambda) - b(\lambda)$; $\mu(\lambda_o) = 0$. Then the equation of the bifurcation set is $\mu(\lambda) = 0$. See Fig. 13a.

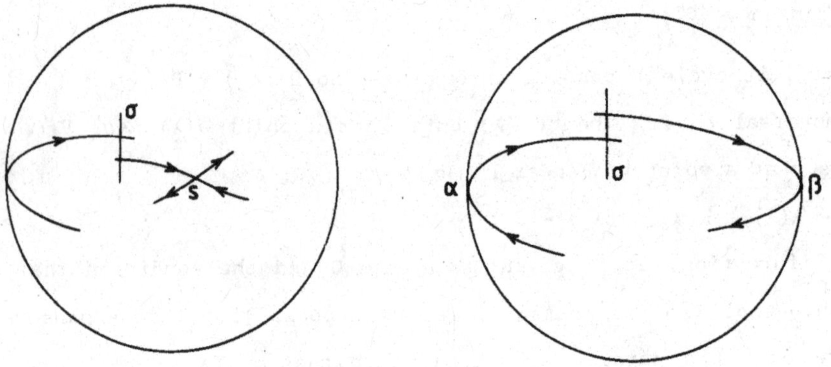

a, Separatrix tangency   b, Double tangency

Figure 13.

8. Double tangency (DT)

Here we suppose that X_{λ_o} has two quadratic inner tangencies α, β with ∂A. Let $\beta(\lambda)$ be the unique prolongation of $\beta = \beta(\lambda_o)$ and $\Lambda^+(\lambda)$, $\Lambda^-(\lambda)$ the half-orbits associated to $\beta(\lambda)$.

We suppose that $\Gamma = \Gamma^+(\lambda_o) = \Lambda^-(\lambda_o)$ (for instance), so that an orbit of X_{λ_o} in A, has a double tangency (in α, β) with ∂A. Let σ be a transversal to Γ and $a(\lambda)$ the point of $\Gamma^+(\lambda) \cap \sigma$ (near $\Gamma \cap \sigma$) and $b(\lambda)$ the point of $\Lambda^-(\lambda) \cap \sigma$ (near $\Gamma \cap \sigma$). Then the bifurcation set is given by $\mu(\lambda) = 0$. See Fig. 13b.

IV.3 : Codimension 2-bifurcations

1. Degenerate Hopf-Takens bifurcation (DH)

Up to orientation, we may suppose that $j^1X_{\lambda_o}(0) = y\frac{\partial}{\partial x} - x\frac{\partial}{\partial y}$. Let $\mu_1(\lambda) \pm i\nu(\lambda)$ be the eigenvalue of the unique singular point of X_λ near 0. ($\mu_1(\lambda_o) = 0$). The family has the following normal form (up to C^∞ equivalence) :

$$X_\lambda \sim y\frac{\partial}{\partial x} - x\frac{\partial}{\partial y} + (\mu_1(\lambda) + \mu_2(\lambda)(x^2+y^2) + \mu_3(\lambda)(x^2+y^2)^2$$

$$+ o((x^2+y^2)^2))(x\frac{\partial}{\partial x} + y\frac{\partial}{\partial y}) \qquad (1)$$

Generic case :

The equations of the bifurcation set are $\mu_1(\lambda) = \mu_2(\lambda) = 0$ and $\mu_3(\lambda_o) \neq 0$

P.H. Case

We suppose that the Hamiltonian is zero at the center. Let us write the h-expansion of the Abelian integral I, up to order 3 at h=0 :

$$I(h,\bar{\lambda}) = \bar{\mu}_1(\bar{\lambda})h + \bar{\mu}_2(\lambda)h^2 + \bar{\mu}_3(\bar{\lambda})h^3 + o(h^3) \qquad (2)$$

Then, the equations of the bifurcation set are given by : $\bar{\mu}_1 = \bar{\mu}_2 = 0$ and $\bar{\mu}_3(\bar{\lambda}_o) \neq 0$.

2. The cuspidal bifurcation (C)

Up to orientation, we may suppose that $j^1X_{\lambda_o}(0) = y\frac{\partial}{\partial y}$ and that the axis Ox is X_λ-invariant. Then we may write :

$$X_\lambda = y(1 + O(\|m\|))\frac{\partial}{\partial y} + (f(x,\lambda) + O(y))\frac{\partial}{\partial x} \qquad (3)$$

The conditions on f are : $f(0,\lambda_o) = \frac{\partial f}{\partial x}(0,\lambda_o) = \frac{\partial^2 f}{\partial x^2}(0,\lambda_o) = 0$ and $\frac{\partial^3 f}{\partial x^3}(0,\lambda_o) \neq 0$. Using the preparation theorem one can find C^∞ functions $\alpha(\lambda)$, $\mu_o(\lambda)$, $\mu_1(\lambda)$ and $u(x,\lambda)$, with $\alpha(\lambda_o) = \mu_o(\lambda_o) = \mu_1(\lambda_o) = 0$ and $u(0,\lambda_o) \neq 0$ such that :

$$f(x+\alpha(\lambda),\lambda) = u(x,\lambda)\,(x^3 + \mu_1(\lambda)\,x + \mu_o(\lambda)) \tag{4}$$

The bifurcation set is given by $\mu_1(\lambda) = \mu_0(\lambda) = 0$ (see [A]).

Remark : Let $f(x,\lambda) = \mu_o'(\lambda) + \mu_1'(\lambda)\,x + O(x^2)$. It is easily seen that the genericity condition (independance of $d\mu_o(\lambda_o)$ and $d\mu_1(\lambda_o)$) is equivalent to the independence of $d\mu_o'(\lambda_o)$ and $d\mu_1'(\lambda_o)$.

3. The Bogdanov-Takens bifurcation (TB)

We may suppose that $j^2X_{\lambda_o}(0) = y\frac{\partial}{\partial x} + (x^2 \pm xy)\frac{\partial}{\partial y}$. Then, the family is C^∞ equivalent to :

$$X_\lambda \sim y\frac{\partial}{\partial x} + (\mu(\lambda) + x^2 + y(\nu(\lambda) \pm x))\frac{\partial}{\partial y} + O(\|m\|^3) + O((\|m\| + \|\lambda\|)^N)$$

(N being an arbitrarily large number).

Then the bifurcation is given by $\mu = \nu = 0$ (see [A], [B1], [B2], [T2]).

4. The degenerate loop (DL)

The vector field X_{λ_o} has a loop Γ through a saddle point $s(\lambda_o)$ where the divergence is zero. Let σ be a transversal to Γ and $s(\lambda)$ the unique singular point of X_λ near $s(\lambda_o)$.

Then $J^1X_\lambda(s(\lambda)) \sim x\frac{\partial}{\partial x} - (1 - \alpha_o(\lambda))\,y\frac{\partial}{\partial y}$ when $\alpha_o(\lambda_o) = 0$.

It is shown in [R] that the return map P_λ on σ has the following expansion (u is a parameter on σ, positive on the side where the return map P_{λ_o} is defined) :

$$P_\lambda(u) = u + \beta_o(\lambda) + \alpha_o(\lambda)\ (u\omega + o(u\omega)) + \beta_1(\lambda)u + o(u) \qquad (5)$$

where $\omega(u,\lambda) = \dfrac{u^{-\alpha_o} - 1}{\alpha_o}$.

Generic case

The equations of the bifurcation set are given by : $\alpha_o = \beta_o = 0$ and $\beta_1(\lambda_o) \neq 0$.

P.H. case

Let Γ be a loop for the Hamiltonian H. We suppose that $\Gamma \subset \{H=0\}$ and that $H > 0$ inside the loop or outside the loop depending on whether the other separatrices of the saddle are outside the loop or inside. For $h > 0$ near 0, the Abelian integral I has the following expansion :

$$I(h,\bar\lambda) = \bar\beta_o(\bar\lambda) + \bar\alpha_o(\bar\lambda)\ h \log h + \bar\beta_1(\bar\lambda)h + o(h) \qquad (6)$$

The equations for the bifurcation set correspond to $\bar\beta_o(\bar\lambda) = \bar\alpha_o(\bar\lambda) = 0$ and $\bar\beta_1(\bar\lambda_o) \neq 0$. This bifurcation has been studied in [DRS].

5. The saddle-node loop of codimension 2 (SNL_2)

At some point (say $0 \in \mathbb{R}^2$) there is a saddle node $sn(\lambda_o)$ and the isolated centre separatrix coincides with one of the hyperbolic separatrices to form a "saddle-node loop of codimension 2" (there are 2 cases). See Fig. 14. In some neighborhood U of 0 we may suppose that the family X_λ is as follows :

$$X_\lambda = -y(1 + O(x)) \frac{\partial}{\partial y} + (\mu(\lambda) + \alpha(\lambda)\, x^2 + O(x^3) + yg(x,y,\lambda)) \frac{\partial}{\partial x} \tag{7}$$

with $\mu(\lambda_o) = 0$ and $\alpha(\lambda_o) > 0$. So in U the axis Oy is the local hyperbolic manifold for X_o and the half-axis Ox $(x > 0)$ is on the isolated centre separatrix for all X_λ.

Now let σ be some transversal to Γ inside U cutting the stable manifold of $sn(\lambda_o)$.

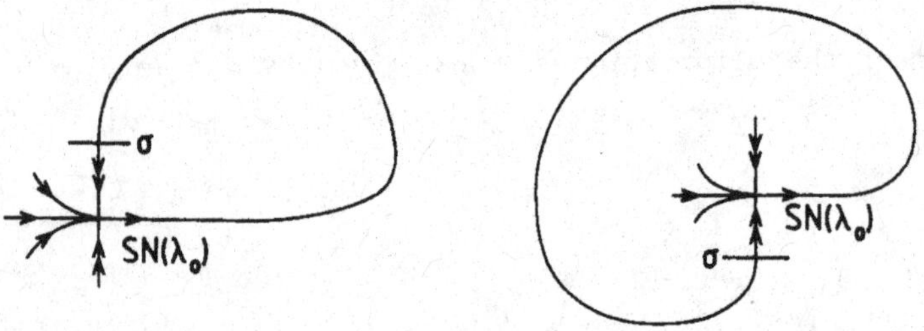

Figure 14.

We use the strong-stable foliation of X_λ to transport the parameter x on Ox to σ.

In this parameter - which we denote by t - the intersection of σ with the stable manifold of the saddle point (for $\mu(\lambda) < 0$) is given by :

$$\mu(\lambda) + \alpha(\lambda)\, t^2 + O(t^3) = 0 \qquad \text{(for some } O(t^3))$$

By an extra coordinate change this can be transformed into $\mu(\lambda)+\alpha(\lambda)s^2 = 0$.

The intersection of σ with the isolated centre separatrix is given by some C^∞ function $\nu = \nu(\lambda)$.

The equations for the bifurcation of codimension 2 are $\nu(\lambda) = \mu(\lambda) = 0$.

The bifurcation diagram reduces to a line of SN : $\mu(\lambda) = 0$, half of which is SNL_1, and a half line of loops L defined by

$$\nu(\lambda) > 0$$

$$\mu(\lambda) + \alpha(\lambda)\, \nu^2(\lambda) = 0$$

A limit cycle exists (here a stable one) everywhere except below L in the half-plane $\mu < 0$ (see figure 15).

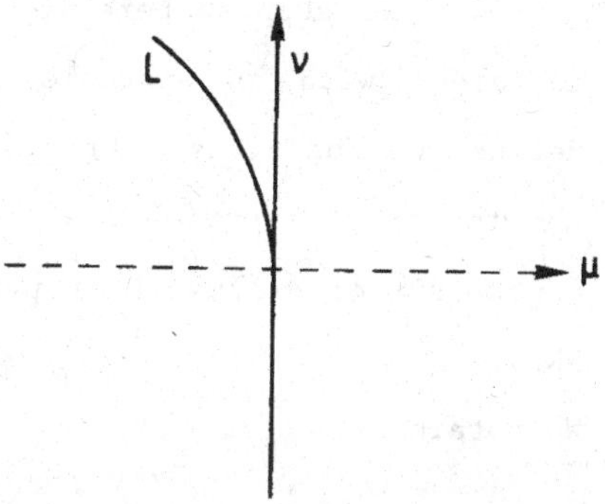

Figure 15a

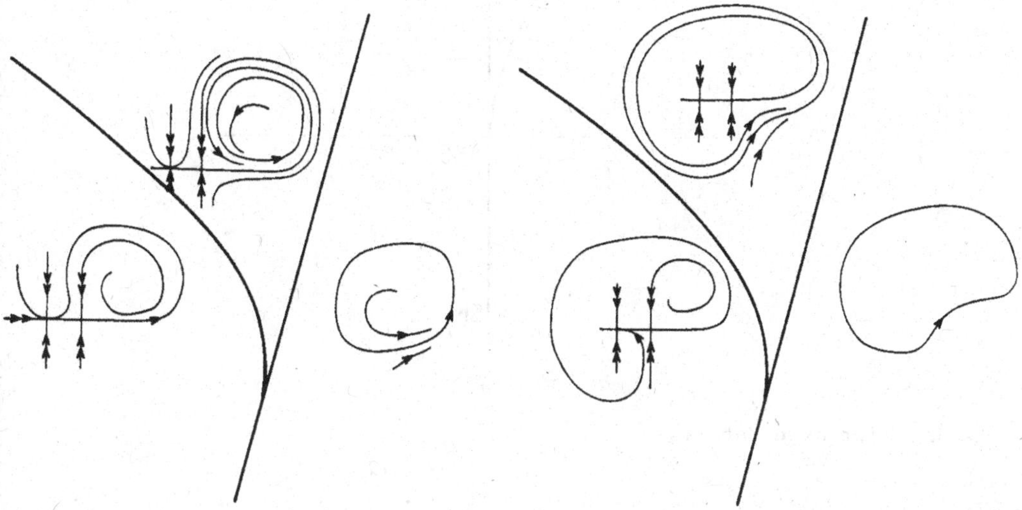

Figure 15b

See [Sc] for more details.

6. Saddle-node connection (SNC)

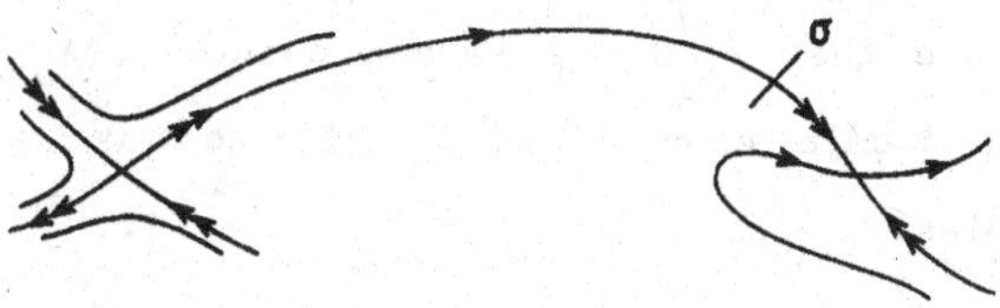

Figure 16

We suppose here that the stable manifold of a saddle-node $sn(\lambda_o)$ coincides with an unstable separatrix of a saddle $s(\lambda_o)$. Again we may define two functions $\mu(\lambda)$ the versal parameter of the saddle-node unfolding and the shift function $\nu(\lambda)$ giving the transverse distance of the unstable separatrix of $s(\lambda)$ with respect to the position of $s(\lambda_o)$ (see 5. concerning SNL).

We obtain :

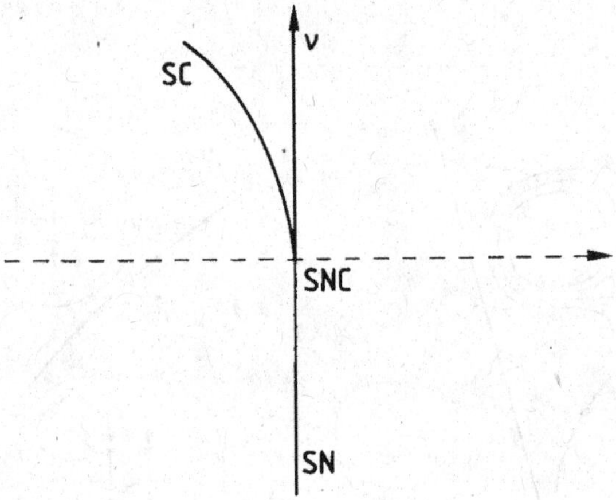

Figure 17

See [Sc] for more details.

The bifurcation point SNC is an end point of a line of separatrix connections SC.

7. Two-saddles cycle (TSC)

We suppose that X_{λ_o} has 2 saddle points $s_1(\lambda_o)$, $s_2(\lambda_o)$ which are connected by two saddle connections Γ_s, Γ_i to make a singular cycle Γ containing 2 saddles.

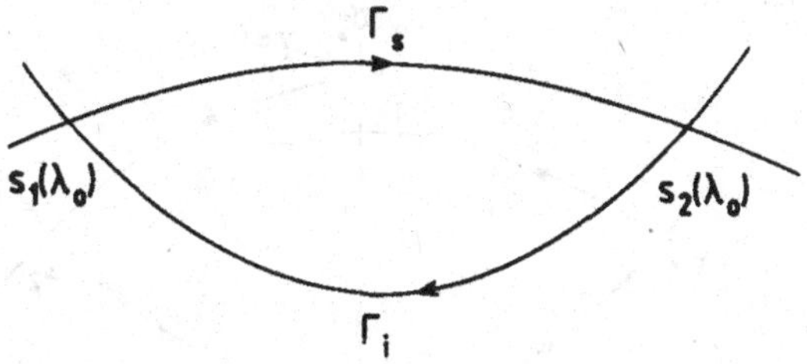

Figure 18

Let $\lambda_1(\lambda_o)$, $-\xi_1(\lambda_o)$ be the eigenvalues at $s_1(\lambda_o)$ and $\lambda_2(\lambda_o)$, $-\xi_2(\lambda_o)$ the eigenvalues at $s_2(\lambda_o)$ ($\lambda_1, \lambda_2, \xi_1, \xi_2 > 0$). We suppose that the two ratios $\frac{\lambda_1}{\xi_1}(\lambda_o)$ and $\frac{\lambda_2}{\xi_2}(\lambda_o)$ are different from 1.

The generic case

Here, we also suppose that $r = \frac{\xi_1}{\lambda_1} \cdot \frac{\xi_2}{\lambda_2} \neq 1$. The singular cyle Γ is attracting if $r > 1$ and expanding if $r < 1$. Up to the orientation we may suppose that we are in the expanding case. Next, up to the order between s_1, s_2, we have two subcases : the strong expanding case $\frac{\xi_1}{\lambda_1} < 1$ and $\frac{\xi_2}{\lambda_2} < 1$ and the weak expanding case : $\frac{\xi_1}{\lambda_1} > 1$ and $\frac{\xi_2}{\lambda_2} < 1$. For our family the last one happens (in the saddle case). So we limit ourselves to the weak expanding case (In fact the other case is easier to study because there is no line of double cycles arriving at the bifurcation point TSC).

Recall the assumptions made on X_{λ_o} : $r_1 = r_1(\lambda_o) = \frac{\xi_1}{\lambda_1}(\lambda_o) > 1$, $r_2 = r_2(\lambda_o) = \frac{\xi_2}{\lambda_2}(\lambda_o) < 1$ and $r_1 r_2 = r(\lambda_o) = r < 1$. For λ near λ_o the saddle points persist in $s_1(\lambda)$, $s_2(\lambda)$, with eigenvalues $\lambda_1(\lambda)$, ... and the ratios $r_1(\lambda)$, $r_2(\lambda)$ and $r(\lambda)$ having the same property ($r_1(\lambda) > 1$, $r_2(\lambda) < 1$, $r(\lambda) < 1$). Taking transversals σ_i, σ_s to Γ_i and Γ_s respectively, we can define shift functions $s(\lambda)$, $i(\lambda)$ on σ_s and σ_i respectively (See figure 19; the orientations of σ_s and σ_i are respectively downwards and upwards).

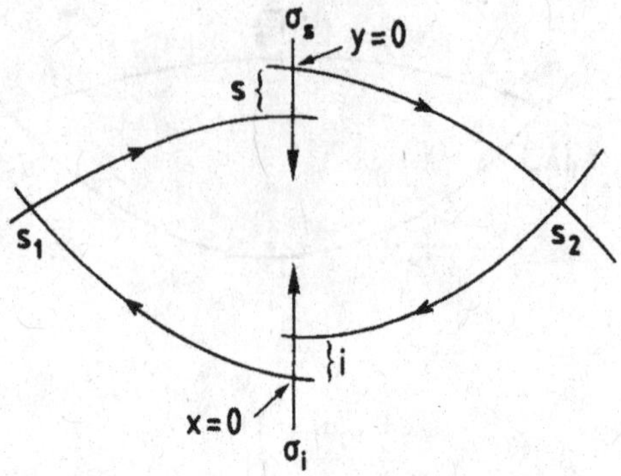

Figure 19

Now, the bifurcation point of cod. 2 is obviously defined by $s(\lambda) = i(\lambda) = 0$. We can describe the bifurcation set near that point in the space (i,s). (See the introduction of the chapter). The lines $\{s=0\}$ and $\{i=0\}$ are lines of saddle connections : SC_s, SC_i (superior and inferior connections). In Part II, we give precise calculations for the other lines of bifurcation. To have a rough idea of them, suppose that the vector field X_λ is linear around each saddle point in the coordinates where s and i are defined. Then the transition map fom σ_i to σ_s is given on the left by : $x \to s + a(\lambda).x^{r_1(\lambda)}$ for some function $a(\lambda) > 0$, and is given on the right, along the orbits of $-X_\lambda$, by : $x \to b(\lambda)\ (-i+x)^{1/r_2(\lambda)}$ (again for some $b(\lambda) > 0$). Then, the equation for the $x \in \sigma_i$ corresponding to limit cycles is :

$$s + a(\lambda)x^{r_1(\lambda)} = b(\lambda)(-i+x)^{1/r_2(\lambda)} \tag{8}$$

($x^{r_1(\lambda)}$ is only defined for $x \geq 0$, and $(-i+x)^{1/r_2(\lambda)}$ for $-i+x \geq 0$). Taking x=0 in this equation, gives an equation for a line L_ℓ of loops on the left (i.e. at s_1). This equation is :

$$L_\ell : s = b(\lambda)\ (-i)^{\frac{1}{r_2(\lambda)}} \text{ for } i \leq 0.$$

Taking the principal term at $\lambda=\lambda_o$, we have the equation :

$$s = b(-i)^{1/r_2},\ i \le 0,\ \text{where } b = b(\lambda_o),\ r_2 = r_2(\lambda_o)$$

In the same way, taking x-i = 0, we find an equation for a line of loops at the right : L_r, whose first order term is :

$$s = -a(i)^{r_1} \text{ for } i \ge 0.\ (a = a(\lambda_o),\ r_1 = r_1(\lambda_o))$$

Next, the equation for double cycles is obtained by adding to (8) its derived equation :

$$r_1(\lambda)a(\lambda)\, x^{r_1(\lambda)-1} \doteq \frac{b(\lambda)}{r_2(\lambda)} (x-i)^{\frac{1}{r_2(\lambda)}-1} \qquad (8')$$

It is easy to eliminate x between (8) and (8') in order to find a line DC whose equation to the first order is again : $s = b(-i)^{1/r_2}$ (the same as that for the line L_ℓ !). However it is easy to prove that the line DC is disjoint from the line L_ℓ and situated as in Fig. 20, using the following arguments : on {i=0,s>0} there are no limit cycles, on {s=0,i<0} there is 1 limit cycle, along L_ℓ there is creation of a second limit cycle for s increasing.

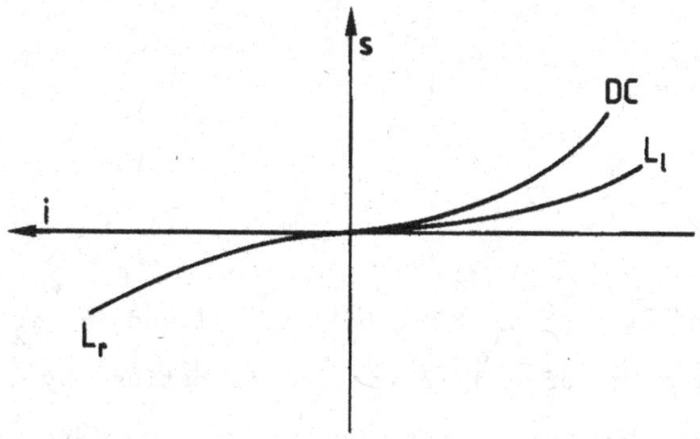

Figure 20

The degenerate case

We again suppose that $r_1 > 1$, $r_2 < 1$, but now we assume that the return mapping along Γ is equal to the identity (for $\lambda = \lambda_o$). This obviously implies : $r_1 r_2 = r = 1$. Contrary to what happens in the case of a simple loop, the vector field X_{λ_o} cannot be C^∞ equivalent to a Hamiltonian one around Γ (because $r_i \neq 1$).

However, to study the codimension 1-bifurcations of saddle connections SC_s, SC_i it will be possible to choose some Hamiltonians which are regular respectively on the interior of Γ_s or Γ_i (but not at the saddle points). See the calculations in VI.B.3.

We need to consider a situation depending on a parameter $\lambda = (\bar{\lambda}, t)$ where for t=0 we have a fixed vector field $\bar{X}$ with the identity as return map. We may suppose that for $\bar{\lambda} \in K$ (some fixed compact set) containing the value $\bar{\lambda}_o$ ($\lambda_o = (\bar{\lambda}_o, 0)$) we have the following expansions for the functions i and s :

$$
\begin{aligned}
i(t,\bar{\lambda}) &= tI(\bar{\lambda}) + o(t) \\
s(t,\bar{\lambda}) &= tS(\bar{\lambda}) + o(t)
\end{aligned}
\tag{9}
$$

with $I(\bar{\lambda}_o) = S(\bar{\lambda}_o) = 0$.

Also, we suppose that the ratio of eigenvalues r_1, r_2 expand as :

$$r_1(\bar{\lambda}, t) = r_1 + t\alpha_1(\bar{\lambda}) + o(t)$$

$$\frac{1}{r_2(\bar{\lambda}, t)} = r_1 + t\alpha_2(\bar{\lambda}) + o(t)$$

The generic hypothesis on X_{λ_o} will be : $r_1 > 1$ and $(\alpha_1 - \alpha_2)(\bar{\lambda}_o) < 0$.

The bifurcation point of cod. 2 for t=0 is defined by : $I(\bar{\lambda}) = S(\bar{\lambda}) = 0$. And so, the genericity of the family is expressed by the independence of these 2 funtions.

Now let $t > 0$ be small enough. Again, we find that there exists a line $SC_s(t)$ which tends to $\{S=0\}$ for $t \to 0$ and a line $SC_i(t)$ which tends to $\{I=0\}$

for $t \to 0$. Also, we find a line $L_\ell(t)$ (loops at the left) whose first term is :

$$S = t^{(r_1-1)} (-I)^{r_1}, \quad I < 0 \tag{10}$$

And a line $L_r(t)$ whose first term is :

$$S = t^{(r_1-1)} (I)^{r_1}, \quad I > 0 \tag{11}$$

These two lines tend towards the axis $\{S=0\}$ for $t \to 0$. Next it is possible to see that there exists a line of double cycles DC(t) which tends towards a well defined position DC for $t \to 0$. This entire description is valid for $\bar{\lambda} \in K$ (a fixed compact subset containing $\bar{\lambda}_o$). In fact in the application to the specific vector field studied in the saddle case, the compact K can be chosen to include the whole line DC up to its other end at the point DH. So, for each t small enough but greater then 0, we find back the situation described in the generic case, on some fixed compact K in the $\bar{\lambda}$-space. For $t=0$ the situation degenerates, with $L_r(t)$, $L_\ell(t)$ and $SC_s(t)$ tending towards the same line $\{S=0\}$.

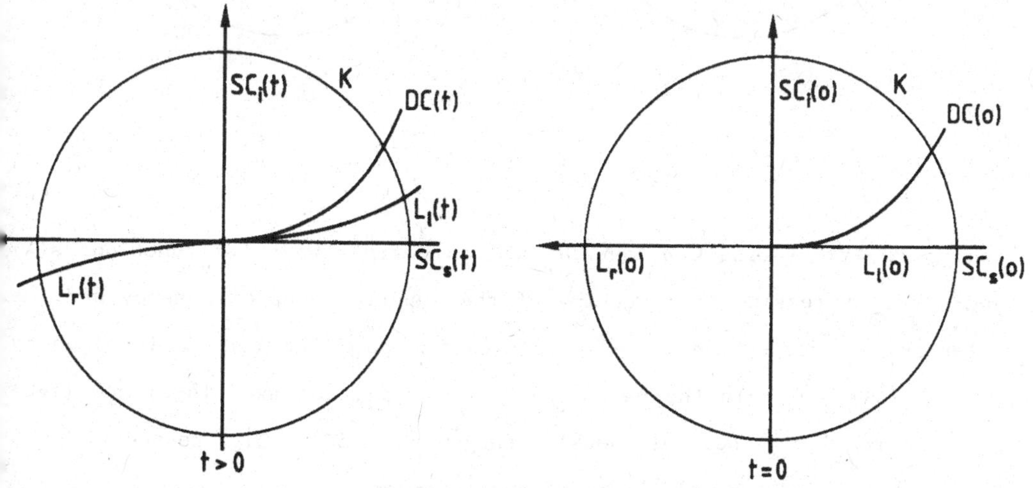

Figure 21

The codimension 2-bifurcations described in 8 to 11 are related to boundary tangencies. Let $\alpha(\lambda_o)$ and possibly $\beta(\lambda_o)$ be two inner quadratic tangencies with the boundary ∂A of some disk A.

8. Twofold cycle tangency (TCT)

A hyperbolic limit cycle Γ of X_{λ_o} passes through $\alpha(\lambda_o)$ and $\beta(\lambda_o)$. This TCT does not really occur in our bifurcation diagrams for the codimension 3 singularities, because of a generic choice of the boundary. It is however almost present and we will say more about it in Part II. Now we give some quick description.

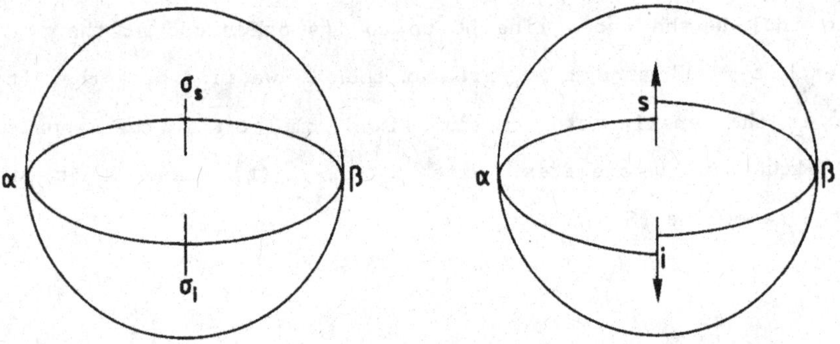

Figure 22

Taking two transversals σ_s and σ_i we can define two shift functions $s(\lambda)$ and $i(\lambda)$, expressing the breaking of the superior double tangency and the inferior one respectively. The bifurcation point of cod. 2 is given by $s(\lambda) = i(\lambda) = 0$. In the parameter space (s,i), the axes $\{s=0\}$ and $\{i=0\}$ are bifurcation lines of double tangency : DT_s, DT_i respectively. Moreover, we find two half-lines of cycle-tangency.

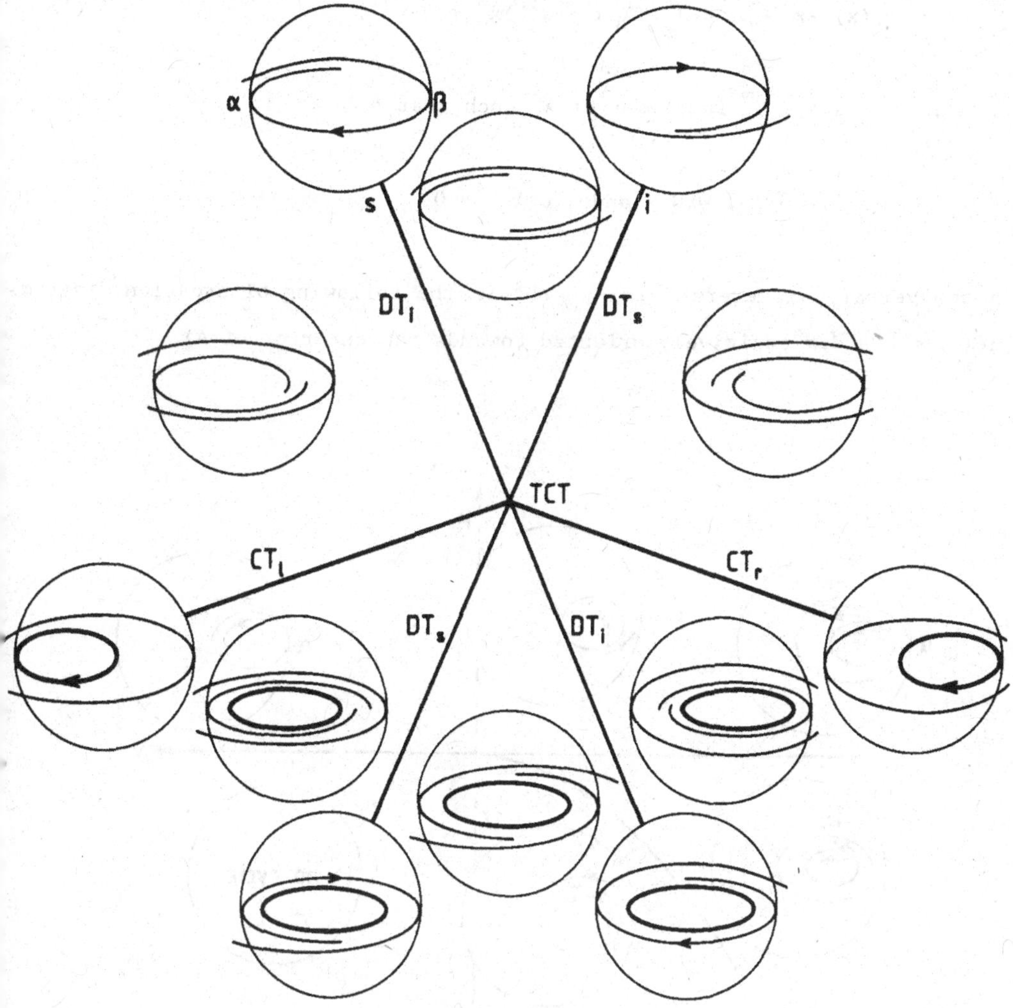

Figure 23

9. <u>Double cycle tangency</u> (DCT)

We suppose that for λ_o a double cycle Γ passes through $\alpha(\lambda_o)$. We take a transversal σ to Γ through $\alpha(\lambda_o)$. Let $P_\lambda(x)$ be the Poincaré-map relative to σ for (x,λ) near $(0,\lambda_o)$ $(\{0\} = \sigma \cap \partial A)$. For defining $P_\lambda(x)$ we need to extend X_λ outside A. Write the x-expansion of P_λ :

$$P_{\lambda}(x) - x = \mu(\lambda) + \nu(\lambda)x + u(\lambda)x^2 + O(x^3), \tag{12}$$

where μ, ν, u are C^{∞} functions of λ, such that :

$$\mu(\lambda_o) = \nu(\lambda_o) = 0 \quad \text{and} \quad u(\lambda_o) \neq 0.$$

In the versal parameters (μ,ν) we obtain the following bifurcation diagram ($u(\lambda_o) < 0$ and σ positively oriented towards the exterior of A).

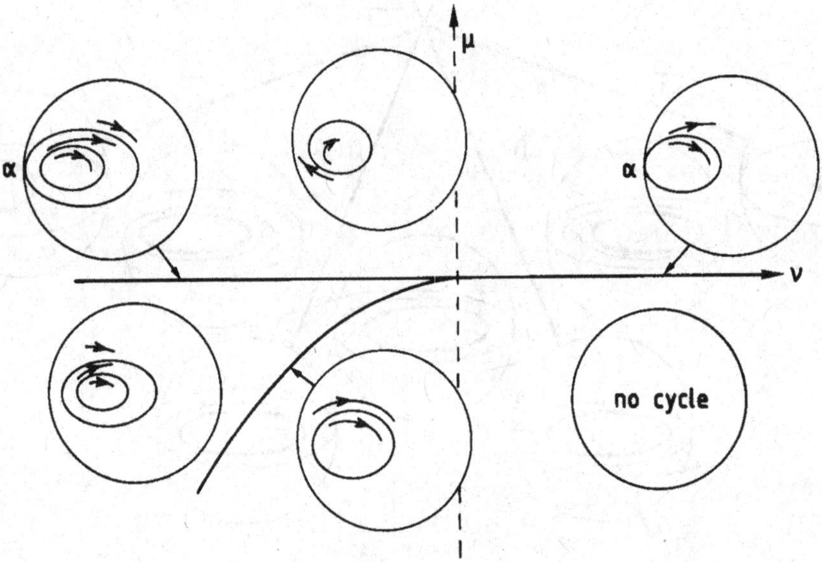

Figure 24

The last bifurcations involve a saddle-node point whose isolated centre separatrix or whose hyperbolic separatrix has a tangency with the boundary. We suppose now that there exist exactly 2 inner quadratic tangencies (and perhaps outer tangencies) α and β. Let us remark that we moreover have possibility of tangency of some orbit in the bundle of non isolated centre orbits (which is an open condition).

10. Hyperbolic separatrix tangency (HST)

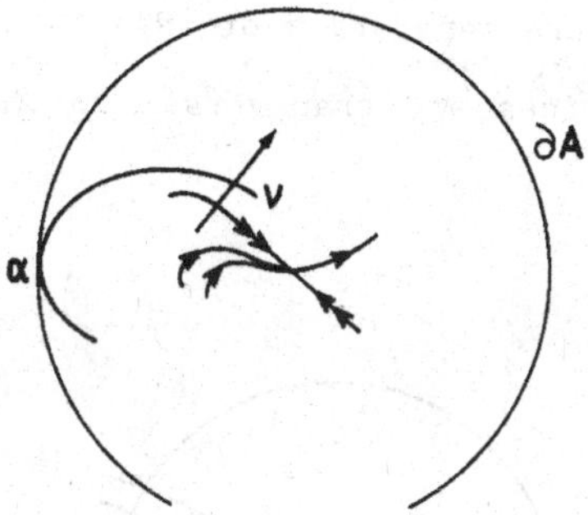

Figure 25

The situation is illustrated in the picture above. Let $\mu(\lambda)$ be the versal parameter of the saddle node unfolding. Taking a transversal σ to the hyperbolic separatrix which is tangent at $\alpha(\lambda_o)$, we define a shift function $\nu(\lambda)$ as it was done before (for the SNL-bifurcation for example). The couple (μ,ν) is the set of versal parameters of the unfolding of the bifurcation point. The bifurcation diagram below shows how a separatrix tangency line ST ends at a point HST on a saddle-node line SN.

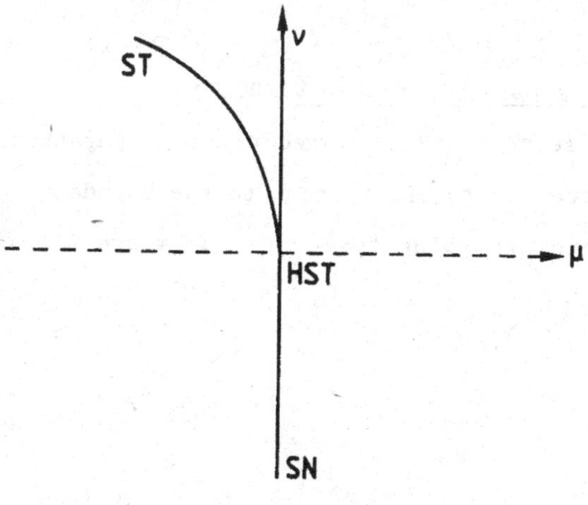

Figure 26

Besides the HST one could expect :

11. Centre separatrix tangency (CST)

The isolated centre separatrix of $SN(\lambda_o)$ is tangent to ∂A in α and all other centre separatrices are transversal to the boundary. (see figure 27).

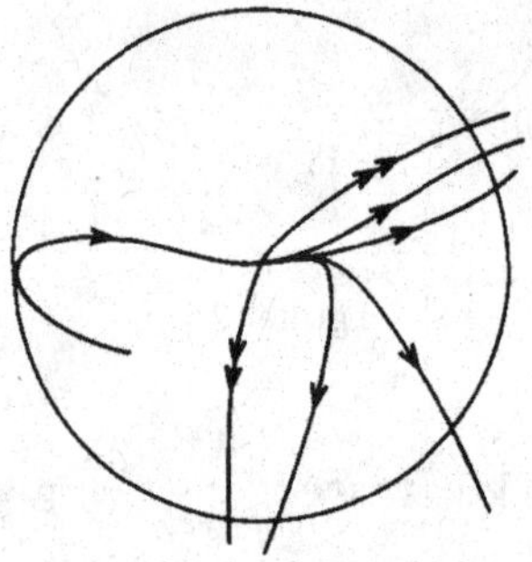

Figure 27

This case does not occur in the families studied here (and its study is rather obvious).

12. Double centre separatrix tangency (DCST)

We suppose that the isolated centre separatrix C and some of the non-isolated centre orbits are tangent to the boundary. We have 3 distinct possibilities which are going to occur. They are illustrated in figure 28 and labeled $DCST_a$, $DCST_b$, $DCST_c$:

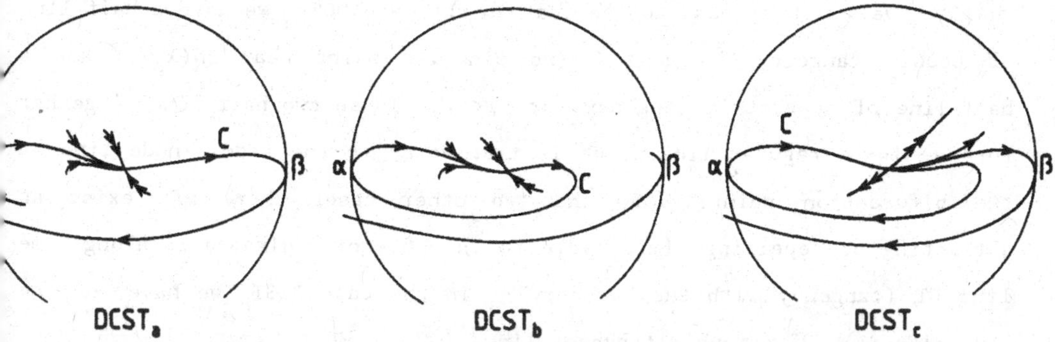

Figure 28

In the case $DCST_a$ the isolated separatrix C is not in the bundle of non-isolated central ones. In the case $DCST_b$ it is in the bundle but there exists just one tangency point (α in the picture) and in the case $DCST_c$, the separatrix C is also in the bundle which has a second tangency (β in the picture). In each case, taking a transversal σ to C we can define a shift function $\nu(\lambda)$ to denote the distance in between a separatrix near C and an orbit tangent to the boundary. We also have the function $\mu(\lambda)$, versal parameter for the unfolding of the saddle-node.

The definition of ν is illustrated in Fig. 29.

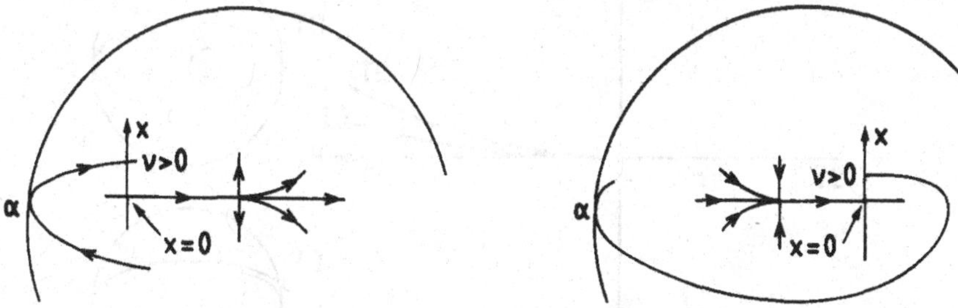

case $DCST_c$ case $DCST_b$ ($DCST_a$ analogously)

Figure 29

Qualitatively, it is easy to see that in the case $DCST_a$ we have a half line of double tangency for $\mu > 0$ (no singular point near $SN(\lambda_o)$), and a half-line of separatrix tangency for $\mu < 0$. These two half lines together form a single regular line which is transverse to the saddle-node line at the bifurcation point. In the two other cases there may exist an attracting or repelling limit cycle for $\mu > 0$, which disappears along some line CT (tangency with the boundary). In the case $DCST_c$ we have, on the same side ($\mu > 0$) a double tangency line.

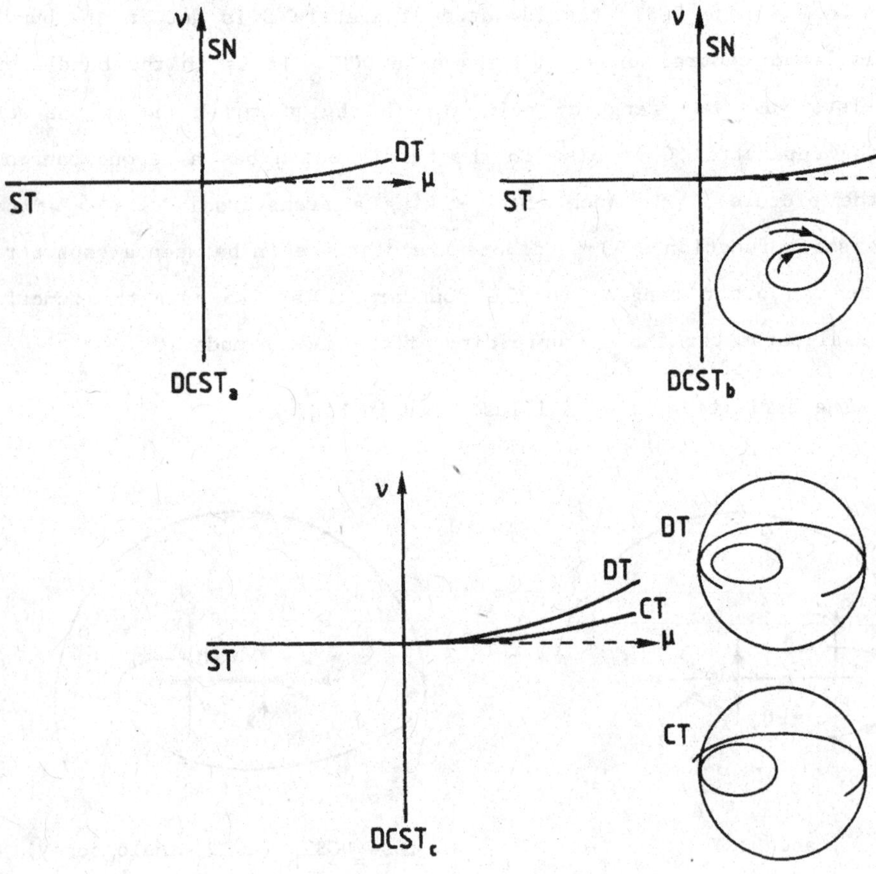

Figure 30

Let us elaborate in more detail the case $DCST_c$, the other ones being easier.

In a sufficiently small neighbourhood of the saddle-node we suppose the family to be

$$y \frac{\partial}{\partial y} + (x^2 + \mu) \frac{\partial}{\partial x}$$

Consider the transversals $\sigma_1 = \{x=-a\}$ and $\sigma_2 = \{x=a\}$ for $a > 0$ sufficiently small. We suppose that the positive orbit of α cuts σ_1 in $y = \nu$, the negative orbit of α cuts σ_2 in $y = \bar{\alpha}(\nu,\mu)$ and the negative orbit of β cuts σ_2 in $y = \bar{\beta}(\nu,\mu)$ with $\bar{\beta}(\nu,\mu) > \bar{\alpha}(\nu,\mu)$; $\bar{\alpha}$ and $\bar{\beta}$ are C^∞, and $-M \leq \bar{\alpha}(\nu,\mu) < \bar{\beta}(\nu,\mu) \leq M$ for ν and μ sufficiently small and $M > 0$.

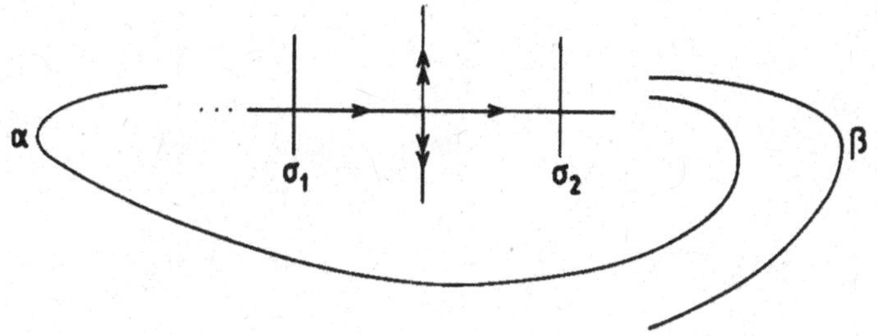

Figure 31

For $\mu > 0$, the orbits cutting σ_1 in y_o will cut σ_2 in $y_o\, e^{(2/\sqrt{\mu})\, \mathrm{arctg}\, a/\sqrt{\mu}}$.

The line CT of cycle tangencies has the equation

$$\nu = \bar{\alpha}(\nu,\mu).\ e^{-(2/\sqrt{\mu})\, \mathrm{arctg}\, a/\sqrt{\mu}} ;$$

the line DT of double tangencies has the equation :

$$\nu = \bar{\beta}(\nu,\mu)\ e^{-(2/\sqrt{\mu})\, \mathrm{arctg}\, a/\sqrt{\mu}}$$

Both are located between the two curves $\nu = \pm M.e^{-(2/\sqrt{\mu})\ \mathrm{arctg}\ a/\sqrt{\mu}}$, and are C^{∞} extensions of the line $ST = \{\nu = 0\}$ for $\mu \leq 0$. The lines CT and DT have infinite contact at $(\nu,\mu) = (0,0)$.

PART II : RESCALINGS AND ANALYTIC TREATMENT

CHAPTER V : ELEMENTARY PROPERTIES

In this chapter we consider simultaneously the saddle, focus and elliptic cases to establish the elementary results concerning the critical points (regions of non-degeneracy, Andronov-Hopf, saddle-node bifurcations, Bogdanov-Takens and degenerate Hopf-Takens bifurcations). Next we establish a "rotational property" for the families under consideration. This property allows a simple study of the loop and saddle connection bifurcation lines. It is also applicable to obtain results on the boundary tangency bifurcation lines in the elliptic case. All these conclusions are obtained directly without appeal to any rescaling.

We introduce next the principal rescaling. Some parts of the bifurcation set obtained before may be recovered in an easier way in the framework of this rescaling. Another of its advantages is to make clear the cone structure at these parts of the bifurcation set.

In what follows we consider a generic family written in the normal form obtained at the end of Chapter III :

$$X_\lambda = y\frac{\partial}{\partial x} + (\epsilon x^3 + \mu_2 x + \mu_1 + y(\nu + b(\lambda)x + x^2 + x^3 h(x,\lambda)) + y^2 Q(x,y,\lambda)))\frac{\partial}{\partial y} \qquad (0)$$

Here we write ϵ for $\epsilon_1 = \pm 1$ and take $\epsilon_2 = +1$ (See the Introduction).

Recall also that $b = b(0) > 0$ and $b \neq 2\sqrt{2}$ for $\epsilon = -1$; $\epsilon = 1$ corresponds to the saddle case and $\epsilon = -1$ to the focus case $(0 < b(0) < 2\sqrt{2})$ or the elliptic case $(b > 2\sqrt{2})$; Q and h are C^∞ functions with $Q(x,y,\lambda) = O((\|m\| + \|\lambda\|)^N)$, where N may be chosen arbitrary large; $\lambda = (\mu_1,\mu_2,\nu) \in \mathbb{R}^3$ is the parameter.

V.A. Location and nature of critical points

The critical points of X_λ are given by $y = 0$ and $\epsilon x^3 + \mu_2 x + \mu_1 = 0 \qquad (1)$

Let SN be the cuspidal surface defined by the zeroes of the discriminant of the first equation : $SN = \{(\mu_1,\mu_2,\nu) \mid 27\mu_1^2 + 4\epsilon\mu_2^3 = 0\}$. The intersection of SN with the sphere $S = \{\mu_1^2 + \mu_2^2 + \nu^2 = \epsilon\}$ gives the lips shaped curve in figures 2, 3, and 4). We verify now that the critical points are non degenerate outside SN.

Let $m_o = (x_o,0)$ be any critical point. Taking $x = x_o + X$, $y = Y$, we calculate the 2-jet of X_λ at m_o :

$$j^2X_\lambda(m_o) = Y\frac{\partial}{\partial X} + (-\mathrm{Det}(x_o,\lambda)X + \mathrm{Tr}(x_o,\lambda)Y + 3\epsilon\, x_oX^2 + (b(\lambda)+2x_o+3x_o^2h(x_o,\lambda)$$
$$+ x_o^3\frac{\partial h}{\partial x}(x_o,\lambda))\,XY + Q(x_o,0,\lambda)Y^2)\frac{\partial}{\partial Y} \tag{2}$$

Where

$$\begin{cases} -\mathrm{Det}(x_o,\lambda) = 3\epsilon x_o^2 + \mu_2 \\ \mathrm{Tr}(x_o,\lambda) = \nu + b(\lambda)x_o + x_o^2 + x_o^3h(x_o,\lambda) \end{cases} \tag{3}$$

In particular, we see that :

$$j^1X_\lambda(m_o) = \begin{pmatrix} 0 & 1 \\ -\mathrm{Det}(x_o,\lambda) & \mathrm{Tr}(x_o,\lambda) \end{pmatrix} \begin{pmatrix} X \\ Y \end{pmatrix} \tag{4}$$

The determinant, $\mathrm{Det}(x_o,\lambda)$, of the 1-jet is non zero when $\lambda \notin SN$ and saddle or focus/node nature of the singular point is given by the sign of $\mathrm{Det}(x_o,\lambda)$, so the problem reduces to the study of the roots of the cubic equation : $\epsilon x^3 + \mu_2 x + \mu_1 = 0$. There exist 3 non-degenerate critical points in the internal region $I = \{27\mu_1^2 + 4\epsilon\mu_2^3 < 0\}$ and 1 non-degenerate point in the external region $E = \{27\mu_1^2 + 4\epsilon\mu_2^3 > 0\}$. The nature of these points can be described as follows :

- In the saddle-case : a focus or node e is located between 2 saddles s_1, s_2 for $\lambda \in I$; a hyperbolic saddle s exists for $\lambda \in E$.

- -In the focus and elliptic cases : a saddle s is located between 2 foci or nodes for $\lambda \in I$; a focus or node exists for $\lambda \in E$

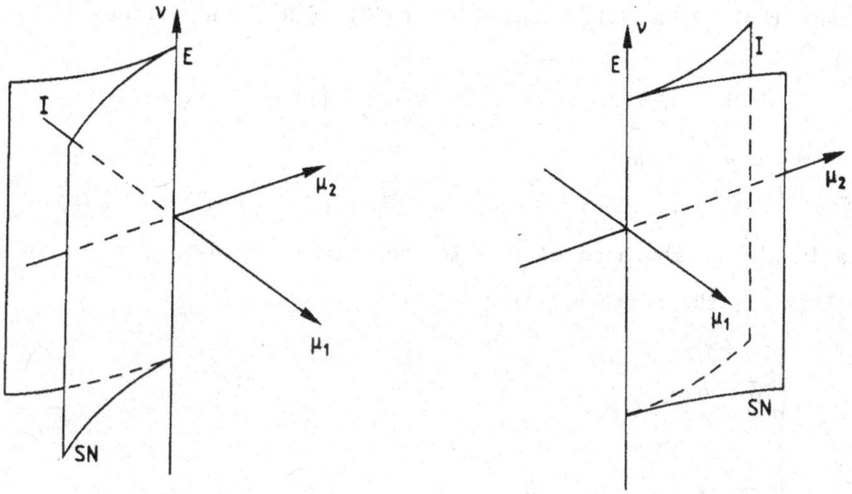

$\epsilon = 1$ (Saddle)

$\epsilon = -1$ (focus, elliptic)

Figure 32

V.B. Location of the Hopf bifurcations of codimensions 1 and 2

The set of Hopf bifurcations of any codimension is contained in the surface T obtained by elimination of x from the 2 equations :

$$\begin{cases} Tr(x,\lambda) = \nu + b(\lambda)x + x^2 + x^3 h(x,\lambda) = 0 \\ \epsilon x^3 + \mu_2 x + \mu_1 = 0 \end{cases} \tag{5}$$

This is the set of values of λ where X_λ has some critical point with vanishing trace.

1. **Basic properties of the surface T**

From

$$\frac{\partial Tr}{\partial \nu}(0,0) = 1 \tag{6}$$

follows that in a neighborhood of $(0,0) \in \mathbb{R}^4$, the surface $\{Tr = 0\}$ is a graph :

$$\nu = \nu(x, \mu_1, \mu_2) \tag{7}$$

Take in the neigborhood of $0 \in \mathbb{R}^4$ the new coordinates $(x, \mu_1, \mu_2, \bar{\nu})$ with $\bar{\nu} = Tr(x,\nu)$, the surface T is :

$$\begin{cases} \bar{\nu} = 0 & (8) \\ \epsilon x^3 + \mu_2 x + \mu_1 = 0 & (9) \end{cases}$$

In the 3-space (x, μ_2, μ_1) equation (9) gives the well known surface $\{\mu_1 = -\mu_2 x - \epsilon x^3\}$.
The critical locus of the projection of this surface on the plane (μ_1, μ_2) is the cusp $\{27\mu_1^2 + 4\epsilon\mu_2^3 = 0\}$.

Equations (5) can also be rewritten as :

$$\begin{cases} f_{10} = \frac{\nu}{b} + x + xR(x,\lambda) = 0 & (10) \\ f_{11} = \epsilon x^3 + \mu_2 x + \mu_1 = 0 & (11) \end{cases}$$

where $R(x,\lambda) = \frac{1}{b}(b(\lambda) - b + x + x^2 h(x,\lambda))$ with $R(0,0) = 0$.

Now, $\frac{\partial f_{10}}{\partial x}(0,0) = 1$. So in a neighborhood of $(0,0)$ the surface $\{f_{10} = 0\}$ is a graph :

$$x = x(\mu_1, \mu_2, \nu) \tag{12}$$

And, as in $(0,0)$: $\frac{\partial f_{10}}{\partial \nu} = \frac{1}{b}$, while $\frac{\partial f_{10}}{\partial \mu_1} = \frac{\partial f_{10}}{\partial \mu_2} = 0$, we find :

$$x = -\frac{\nu}{b} + H(\nu, \mu_1, \mu_2) \tag{13}$$

with $H(\lambda) = 0(\| \lambda \|^2)$

The mapping $(\mu_1, \mu_2, \nu) \to (\mu_1, \mu_2, -\frac{\nu}{b} + H(\lambda))$ is a local diffeomorphism around 0 and the inverse image of (9) gives in the (μ_1, μ_2, ν)-space the surface T :

$$\epsilon(-\frac{\nu}{b} + H(\lambda))^3 + \mu_2(-\frac{\nu}{b} + H(\lambda)) + \mu_1 = 0 \tag{14}$$

illustrated in Figs. 33

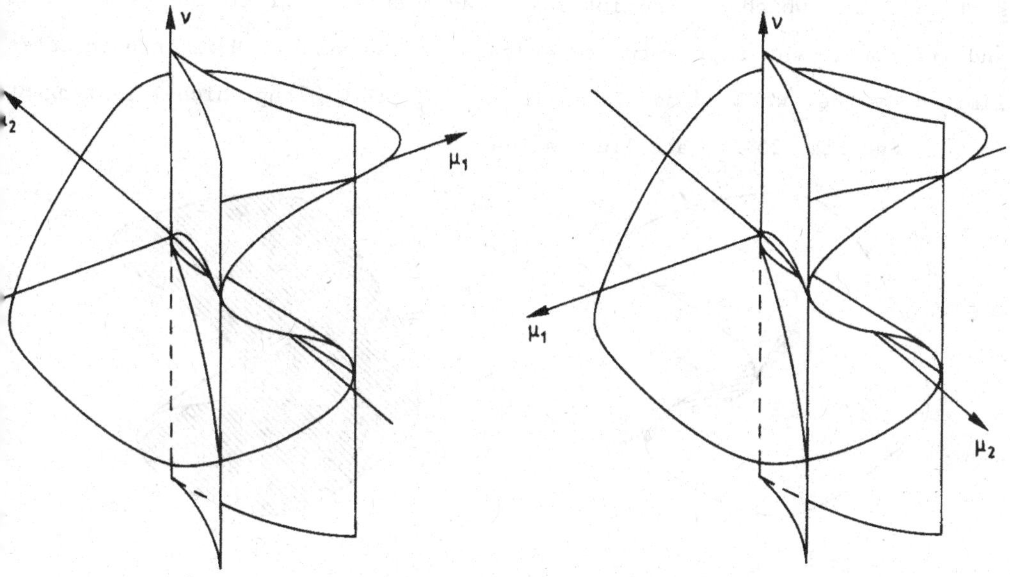

$\epsilon = 1$ $\epsilon = -1$

Surface T

Figure 33

The shape of this surface is influenced by b as illustrated in Fig. 34

small b

large b

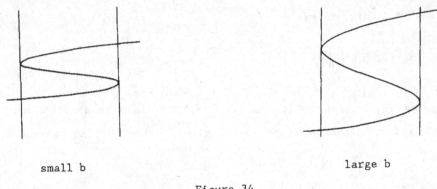

Figure 34

To finish we have to restrict the condition $(Tr(x,\lambda) = 0)$ of (5) to the focal points. Using paragraph A we obtain what we call now the <u>Hopf surface H</u> and which is the internal middle sheet of T in the saddle case and its complement in the two other cases. The surface H is in both cases limited by the two Bogdanov-Takens lines TB_{ℓ}, TB_r, along which T is tangent to SN. See Fig. 35 for an illustration.

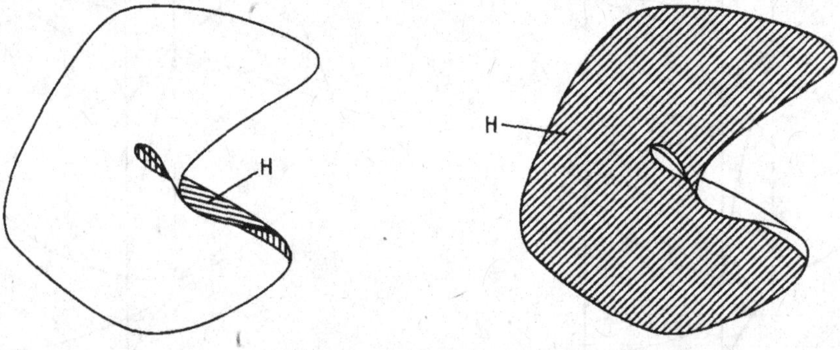

$\epsilon = 1$

$\epsilon = -1$

Figure 35

2. Position of the line of Hopf-bifurcations of codimension greater than 1

The analysis will actually show that there is just one line of such Hopf bifurcations.

We start again from equations (5), and suppose (a,0) is a critical point. In order to calculate the first Lyapounov coefficient at this point, we perform the translation :

$$(x', y) = (x-a, y) \tag{15}$$

Omitting the primes, we obtain :

$$y \frac{\partial}{\partial x} + [(\epsilon x^2 + 3\epsilon\, ax + (3\epsilon a^2 + \mu_2))\, x + y\, (\nu + b(\lambda)(x+a) + x^2 + 2ax + a^2 + a^3 h(a,\lambda)$$

$$+ ((x+a)^3 h(x+a,\lambda) - a^3 h(a,\lambda))) + y^2 Q(x+a,y,\lambda)] \frac{\partial}{\partial y} \tag{16}$$

which, because of $\nu + b(\lambda)a + a^2 + a^3 h(a,\lambda) = 0$ gives :

$$y \frac{\partial}{\partial x} + [(\epsilon x^2 + 3\epsilon\, ax + (3\epsilon a^2 + \mu_2))\, x + y((b(\lambda)+2a)x + x^2$$

$$+ ((x+a)^3\, h(x+a,\lambda) - a^3\, h(a,\lambda))) + y^2\, Q(x+a,y,\lambda)] \frac{\partial}{\partial y} \tag{17}$$

As the singularity must be a focus (Det $(a,\lambda) > 0$) we have :

$$3\epsilon a^2 + \mu_2 < 0 \tag{18}$$

The coordinate change :

$$\begin{cases} x = x' \\ y = (-(3\epsilon a^2 + \mu_2))^{\frac{1}{2}} y' \end{cases} \tag{19}$$

gives, omitting the primes and writing $\Delta = \mathrm{Det}(a,\lambda) = -(3\epsilon a^2 + \mu_2))$:

$$\sqrt{\Delta}\, y \frac{\partial}{\partial x} + [\frac{1}{\sqrt{\Delta}} (\epsilon x^2 + 3\epsilon\, ax - \Delta)x + y((b(\lambda)+2a)x + x^2 + ((x+a)^3\, h(x+a,\lambda)$$

$$- a^3 h(a,\lambda))) + \sqrt{\Delta}\, y^2\, Q(x+a,\, \sqrt{\Delta}\, y,\, \lambda)] \frac{\partial}{\partial y} \tag{20}$$

Multiplying this expression by $\frac{1}{\sqrt{\Delta}}$ we find :

$$y\frac{\partial}{\partial x} + [\frac{\epsilon x^3}{\Delta} + \frac{3\epsilon a x^2}{\Delta} - x + y(\frac{b(\lambda)+2a}{\sqrt{\Delta}}x + \frac{x^2}{\sqrt{\Delta}} + \frac{1}{\sqrt{\Delta}}((x+a)^3 h(x+a,\lambda)$$

$$- a^3 h(a,\lambda))) + y^2 Q(x+a, \sqrt{\Delta}y, \lambda)]\frac{\partial}{\partial y} \tag{21}$$

As the 1-jet is $y\frac{\partial}{\partial x} - x\frac{\partial}{\partial y}$, the formula in Chapter IV.2.1 gives as first Lyapounov coefficient

$$-\frac{1}{\sqrt{\Delta}}(1 + \bar{A} + 3\Delta\frac{\partial Q}{\partial y}(a,0,\lambda)) + \frac{b(\lambda)+2a+\bar{B}}{\sqrt{\Delta}}(\frac{-3\epsilon a - \Delta Q(a,0,\lambda)}{\Delta}) \tag{22}$$

Where $\bar{A}$ = coefficient of x^2 in $((x+a)^3\ h(x+a,\lambda)\ -\ a^3 h(a,\lambda))$

$\bar{B}$ = coefficient of x in $((x+a)^3\ h(x+a,\lambda)\ -\ a^3 h(a,\lambda))$

Clearly, $\bar{A}$, $\bar{B}$ are of the form : $\bar{A} = aA$, $\bar{B} = a^2B$. The expression (22) has the same sign as :

$$[-3\epsilon a + (3\epsilon a^2 + \mu_2)\ Q(a,0,\lambda)][b(\lambda) + 2a + a^2 B]$$

$$+ [3\epsilon a^2 + \mu_2][1 + aA - 3(3\epsilon a^2 + \mu_2)\frac{\partial Q}{\partial y}(a,0,\lambda)] \tag{23}$$

The Hopf bifurcations of codimension > 1 can only appear when this expression (23) is zero. If we consider the mapping $\Phi : \mathbb{R}^4 \to \mathbb{R}^3$, $(a,\nu,\mu_1,\mu_2) \to \Phi(a,\nu,\mu_1,\mu_2)$ whose components are the expressions in (5) and (23), then we see that :

$$D_\lambda\Phi(0) = P_\lambda$$

where P_λ denotes the projection on the parameter space. Therefore, locally around 0, the solution of $\{\Phi = 0\}$ is a line $\nu = \nu(a)$, $\mu_1 = \mu_1(a)$, $\mu_2 = \mu_2(a)$. From (5) it follows that ν and μ_1 are O(a). Therefore $Q(a,0,\lambda) = O(a) + O(\mu_2)$.

So, from (23) it follows :

$$[0(\mu_2^2) + 0(a)][b + 0(a) + 0(\mu_2)] + [\mu_2 + 0(a)][1+ 0(a) + 0(\mu_2)] = 0 \qquad (24)$$

Hence μ_2 is $0(a)$ and $Q(a,0,\lambda) = 0(a^N) = 0(a^2)$

$$\frac{\partial Q}{\partial y}(a,0,\lambda) = 0(a^{N-1}) = 0(a^2)$$

Inserting these expressions into (23) we obtain that

$$(-3\epsilon a + 0(a^2))(b + 0(a)) + (3\epsilon a^2 + \mu_2)(1 + 0(a)) = 0 \qquad (25)$$

From (5) and (25) we get :

$$\begin{cases} \mu_2 = 3\epsilon ba + 0(a^2) \\ \mu_1 = -3\epsilon ba^2 + 0(a^3) \\ \nu = -ba + 0(a^2) \end{cases} \qquad (26)$$

From (26) and (18) we see that $\mu_2 \sim 3\epsilon ba$ while $\mu_2 < -3\epsilon a^2$. For small neighborhoods of 0, this is only possible if $a > 0$, in case $\epsilon = -1$, and if $a < 0$, in case $\epsilon = 1$. In all cases we get $\mu_2 < 0$. Writing $a = t^2$ in the case $\epsilon = -1$ and $a = -t^2$ in the case $\epsilon = 1$, this gives the following parametrization for the line DH of Hopf bifurcation of codimension > 1 :

Case $\epsilon = -1$ (focus, elliptic) :

$$\begin{cases} \mu_2 = -3bt^2 + 0(t^4) \\ \mu_1 = 3bt^4 + 0(t^6) \\ \nu = -bt^2 + 0(t^4) \end{cases} \qquad (27)$$

Case $\epsilon = 1$ (saddle) :

$$\begin{cases} \mu_2 = -3bt^2 + 0(t^4) \\ \mu_1 = -3bt^4 + 0(t^6) \\ \nu = bt^2 + 0(t^4) \end{cases} \qquad (28)$$

Notice that we have not yet proved that the line DH is a line of generic codimension 2 Hopf-Takens bifurcations. To prove this point, we must verify that the second Lyapounov exponent does not vanish and check the genericity (transversality conditions) for the family. In fact, we will proceed differently in Chapter VI, using the perturbation of a Hamiltonian (central rescaling).

Finally, it is easy to see that at each $\lambda \in$ H-DH $\cup$ $TB_\ell \cup TB_r$ the family is a generic Andronov-Hopf bifurcation.
We already know that for such λ, the 1st Lyapounov coefficient is non zero. The trace at the corresponding focus $(a(\lambda), 0)$ is :

$$Tr(a,\lambda) = \nu + b(\lambda)a + a^2 + a^3\ h(a,\lambda). \tag{29}$$

Locally, we can replace the parameter (μ_1,μ_2,ν) by (a,μ_2,ν) using the formula $\mu_1 = -\mu_2 a - \epsilon a^3$ (which gives a diffeomorphism when $\mu_2 \neq 0$). Then $\frac{\partial Tr}{\partial \nu} = 1 + a\frac{\partial b}{\partial \nu} + a^3\frac{\partial h}{\partial \nu}$ is non zero if a is small enough; which proves the transversality of the codimension one Hopf bifurcation.

V.C. Bifurcations along the set SN

Recall that the vector field X_λ has a degenerate singular point for $\lambda \in SN = \{27\mu_1^2 + 4\epsilon\mu_2^3 = 0\}$. Let $(x_o,0)$ be this point. We have :

$$\begin{cases} \epsilon x_o^3 + \mu_2 x_o + \mu_1 = 0 \\ 3\epsilon x_o^2 + \mu_2 = 0 \end{cases} \tag{30}$$

It is useful to parametrize SN by x_o and ν. From (30) we obtain :

$$\mu_1 = 2\epsilon x_o^3,\quad \mu_2 = -3\epsilon x_o^2,\quad \nu = \nu \tag{31}$$

1. The Bogdanov-Takens bifurcation line : TB

The trace at the point $(x_o,0)$ is given by :

$$Tr(x_o,\lambda) = \nu + b(\lambda)\, x_o + x_o^2 + x_o^3\, h(x_o,\lambda) \qquad (32)$$

The point is nilpotent if $Tr(x_o,\lambda) = 0$ (33)

Notice that $\frac{\partial Tr}{\partial \nu} = 1 + 0(x_o)$. So equation (33) is inversible in ν if x_o is small enough. From this, it follows that $\nu = 0(x_o)$, that $b(\lambda) = b + 0(x_o)$ and finally that

$$\nu = -bx_o + 0(x_o^2) \qquad (34)$$

Therefore there exists a regular line TB on SN, parameterized by x_o small enough : $\mu_1 = 2\epsilon x_o^3$, $\mu_2 = -3\epsilon x_o^2$, $\nu = -bx_o + 0(x_o^2)$ which contains all the λ's where X_λ has a nilpotent critical point. This line passes through $0 \in \mathbb{R}^3$ and splits into 2 parts : TB_r and TB_ℓ, depending on the sign of x_o.

Let $\lambda_o \in TB - \{0\}$, $\lambda_o = (\mu_1^o, \mu_2^o, \nu^o)$. Let :

$$\mu_1 = \mu_1^o + M_1, \quad \mu_2 = \mu_2^o + M_2, \quad \nu = \nu^o + N, \; x = x_o + X, \; y = Y$$

and $\Lambda = (M_1, M_2, N)$. Take $X, M_1, M_2, N \in [-x_o, x_o]$.

We develop now the family X_λ in the coordinates X, Y and parameter Λ; x_o enters also in the formula, regarded as an arbitrarily small extra parameter

$$X_{\lambda_o+\Lambda} = Y\frac{\partial}{\partial X} + (\epsilon x_o^3 + \mu_2 x_o + \mu_1 + (3\epsilon x_o^2 + \mu_2)X + 3\epsilon x_o X^2 + \epsilon X^3$$

$$+ \psi(X,\lambda)X^2Y + Tr(x_o,\lambda)Y + (b(\lambda) + 2x_o + \frac{\partial}{\partial x}(x^3h)_o)XY + \Phi Y^2)\frac{\partial}{\partial Y} \qquad (35)$$

where $\Phi = \Phi(X,Y,\Lambda,x_o)$ is $O(x_o^D) + O(Y^D)$ for an arbitrarily large D. If we take into acount that : $\epsilon x_o^3 + \mu_2^o x_o + \mu_1^o = 0$, $3\epsilon x_o^2 + \mu_2^o = 0$, $Tr(x_o,\lambda_o) = 0$, we have :

$$X_{\lambda_o+\Lambda} = Y\frac{\partial}{\partial X} + ((M_2 x_o + M_1) + M_2 X + 3\epsilon x_o X^2 + \epsilon X^3 + (N+H)Y + (b+L)XY$$

$$+ \psi(X,\lambda)X^2 Y + \Phi Y^2)\frac{\partial}{\partial Y} \tag{36}$$

where H, L = $O(x_o)$.

We can reduce $X_{\lambda_o+\Lambda}$ to the Takens normal form by a Λ-dependent diffeomorphism $G_\lambda = Id + O(M^2)\, O(x_o^D)$, $M = (X,Y)$. We find :

$$X_{\lambda_o+\Lambda} = Y\frac{\partial}{\partial X} + (F(X,\Lambda) + YG(X,\Lambda) + Y^2 Q(X,Y,\Lambda))\frac{\partial}{\partial Y} \tag{37}$$

where $F = M_2\, x_o + M_1 + (M_2 + O(x_o^D)\, O(\Lambda))X + (3\epsilon x_o + O(x_o^D))X^2 + O(X^3)$

$$G = (N + H) + (b+L+O(x_o^D))\, X + O(X^2)$$

and $Q = O((\|M\| + \|\Lambda\|)^D)$.

Finally we can suppress the term $(M_2 + O(x_o^D)\, O(\Lambda))X$ by an X-translation of the type : $X \to X + \frac{1}{x_o}(O(M_2) + O(x_o^D)\, O(\Lambda))$ which gives :

$$X_{\lambda_o+\Lambda} \sim Y\frac{\partial}{\partial X} + (M_2 x_o + M_1 + \frac{1}{x_o}(O(M_2) + O(x_o^D)\, O(\Lambda)) + (3\epsilon x_o + O(x_o^D))X^2$$

$$+ Y(N+H+\frac{1}{x_o}(O(M_2) + O(x_o^D)\, O(\Lambda)) + (b+L+O(x_o^D))X)$$

$$+ O(\|M\|^3) + Y^2\, O((\|M\| + \|\Lambda\|)^D))\frac{\partial}{\partial Y} \tag{38}$$

which for each x_o separately (and sufficiently small) has the form of a generic Λ-Bogdanov-Takens bifurcation.

Remark. Here, $O((\|M\| + \|\Lambda\|)^D)$ is not meant to be valid uniformly in x_o.

2. The saddle-node bifurcations (of codimension 1 and 2)

We suppose now that $\lambda_o \in SN\backslash TB$, so that $Tr(x_o,\lambda_o) \neq 0$. Begin again with formula (35) :

$$X_{\lambda_o+\Lambda} = Y\frac{\partial}{\partial X} + (a(\Lambda) + b(\Lambda)X + c(\Lambda)Y(1 + O(\|M\|)) + 3\epsilon x_o X^2 + \epsilon X^3)\frac{\partial}{\partial Y} \tag{39}$$

where : $a(\Lambda) = M_1 + x_o M_2$, $b(\Lambda) = M_2$, $c(\Lambda) = Tr(x_o,\lambda)$ $(c(0) = Tr(x_o,\lambda_o) \neq 0)$

Let c_i and c_s be respectively the negative and positive half-axes $o\nu$.

<u>Lemma 1</u> : Let $m_o = (x_o,0)$. then

$$j^2X_{\lambda_o}(m_o) \sim c(0)\,Y\frac{\partial}{\partial Y} - \frac{3\epsilon x_o}{c(o)}X^2\frac{\partial}{\partial X} \quad \text{for } \lambda_o \in SN \setminus TB \text{ and } \lambda_o \notin c_i \cup c_s \text{ ;}$$

$$j^3X_{\lambda_o}(m_o) \sim c(o)\,Y\frac{\partial}{\partial Y} - \frac{\epsilon}{c(o)}X^3\frac{\partial}{\partial X} \quad \text{for } \lambda_o \in c_i \cup c_s - \{0\}.$$

<u>Proof</u>. Obviously, for $\lambda_o \in SN \setminus TB$: $j^1X_{\lambda_o}(m_o) = Y\frac{\partial}{\partial X} + c(o)\,Y\frac{\partial}{\partial Y}$.

The central axis is OX. Then, each central manifold W (which is C^∞) has an expression :

$$W : Y = \psi(X) = KX^2 + O(X^3) \tag{40}$$

Obviously, the restriction of X_{λ_o} to W has the following orbit equation :

$$\dot{X} = \psi(X) \tag{41}$$

To find the coefficient K, we write that W is invariant by X_{λ_o}, i.e. at the point $(X, \psi(X))$ the tangent vector to W has the same direction as $X_{\lambda_o}(X,\psi(X))$:

$$\frac{\dot{Y}}{\dot{X}}(X,\psi) = \frac{d\psi}{dX} \tag{42}$$

This equation gives : $$\frac{c(o)KX^2 + 3\epsilon x_o X^2 + O(X^3)}{KX^2 + O(X^3)} = 2KX + O(X^2) \tag{43}$$

This implies that : $c(o)K + 3\epsilon x_o = 0$ and the first result follows.

If now $\lambda_o \in c_i \cup c_s - \{0\}$, it follows easily that the central manifolds W are of the form :

$$Y = \psi(X) = KX^3 + O(X^4) \tag{44}$$

Again, applying (42) we obtain :

$$\frac{c(o)\,K\,X^3 + \epsilon X^3 + O(X^4)}{KX^3 + O(X^4)} = 3KX^2 + O(X^3), \tag{45}$$

and the desired result follows.

Lemma 2. The family $X_{\lambda_o+\Lambda}$ is a generic (codimension 1) saddle node bifurcation for $\lambda_o \in SN \setminus (TB \cup c_i \cup c_s)$ and a generic (codimension 2) cuspidal bifurcation for $\lambda_o \in c_i \cup c_s - \{0\}$.

Proof. Lemma 1 established that the vector field X_{λ_o} has the correct form.

It suffices to prove now that the Λ-family is generic. We proceed for the family as we did for the vector field X_{λ_o}. Suppose that $W^\Lambda : Y = \Phi(X,\Lambda)$ is an equation for a central manifold for the family. The restriction of $X_{\lambda_o+\Lambda}$ to W^Λ has the orbit equation :

$\dot{X} = \Phi(X,\Lambda)$, where W^Λ is parametrized by X.

Consider, to begin with, $\lambda_o \in S_N \setminus (TB \cup c_i \cup c_s)$. We look at

$$\Phi(X,\Lambda) = A(\Lambda) + B(\Lambda)X + K(\Lambda)X^2 + O(X^3) \tag{46}$$

with $A(0) = B(0) = 0$ and $K(0) = -\frac{3\epsilon x_o}{c(o)}$ calculated above. We obtain the first order terms of A from equation (42) applied to X_λ.

We have : $a(\Lambda) + c(\Lambda)\ A(\Lambda) = o(\Lambda)$ which implies :

$$A(\Lambda) = -\frac{a(\Lambda)}{c(\Lambda)} + o(\Lambda). \tag{47}$$

Obviously $da(0) \neq 0$ and so $dA(0) \neq 0$, which is the transversality condition.

If $\lambda_o \in c_i \cup c_s$, we look for :

$$\Phi(X,\Lambda) = A(\Lambda) + B(\Lambda)X + C(\Lambda)X^2 + K(\Lambda)X^3 + O(X^4) \tag{48}$$

Again, formula (42) applied to the family $X_{\lambda_o+\Lambda}$ gives :

$$\begin{cases} a(\Lambda) + c(\Lambda)\ A(\Lambda) = o(\Lambda) \\ b(\Lambda) + c(\Lambda)\ B(\Lambda) = o(\Lambda) \end{cases} \tag{49}$$

And the independence of $dA(0)$, $dB(0)$ follows from the obvious independence of $da(0)$, $db(o)$. This proves the genericity (see remark in III.B.2).

V.D. Rotational property with respect to the parameter ν

1. The property

Take a fixed point $m = (x,y)$ and a fixed value $\mu = (\mu_1,\mu_2)$. We want to compare the values of the vector field $X_\lambda(m)$ for $\lambda_o=(\mu,\nu_o)$ and $\lambda_1 = (\mu,\nu_1)$. We have :

$$X_{\lambda_1} - X_{\lambda_o} = (\nu_1-\nu_o)\ (1 + \frac{\partial b}{\partial \nu}(\lambda')x + x^3\frac{\partial h}{\partial \nu}(x,\lambda'') + y\frac{\partial Q}{\partial \nu}(x,y,\lambda'''))y\ \frac{\partial}{\partial y} \tag{50}$$

where λ', λ'', $\lambda''' \in [\lambda_o,\lambda_1]$, possibly dependent on x or (x,y).

Let

$$\Phi(x,y,\lambda',\lambda'',\lambda''') = \frac{\partial b}{\partial \nu}(\lambda')x + \frac{\partial h}{\partial \nu}(x,\lambda'')x^3 + \frac{\partial Q}{\partial \nu}(x,y,\lambda''')y \tag{51}$$

We can choose the compact neighbourhood AxB for the family X_λ such that

$$|\Phi(x,y,\lambda',\lambda'',\lambda''')| \le 1/2 \quad \text{for } (x,y) \in A \text{ and } \lambda', \lambda'', \lambda''' \in B \tag{52}$$

Then

$$(\nu_1-\nu_o)(1+\Phi) \ge \frac{1}{2}(\nu_1-\nu_o) \quad \text{if} \quad \nu_1 \ge \nu_o \tag{53}$$

This amounts to a rotational property of the family X_λ in function of the parameter ν. More precisely if we compute :

$X_{\lambda_o} \wedge (X_{\lambda_1} - X_{\lambda_o}) = \langle X^\perp_{\lambda_o}, X_{\lambda_1} - X_{\lambda_o} \rangle$ where $X^\perp_{\lambda_o}$ is X_{λ_o} rotated by angle $+\pi/2$, we have :

$$\langle X^\perp_{\lambda_o}, X_{\lambda_1} - X_{\lambda_o} \rangle = \langle y\frac{\partial}{\partial y}, X_{\lambda_1} - X_{\lambda_o} \rangle = (\nu_1-\nu_o)y^2(1+\Phi) \tag{54}$$

This means that for $\nu_1 > \nu_o$ and for $y \ne 0$ one passes from the oriented direction of X_{λ_o} to that of X_{λ_1} by a direct rotation of positive angle. The speed of this rotation is given by :

$$\frac{\partial}{\partial \nu} \langle X^\perp_{\lambda_o}, X_\lambda \rangle \Big|_{\nu=\nu_o} = y^2(1+\Phi) \ge \frac{1}{2}y^2 \tag{55}$$

Position of X_{λ_1}, $\lambda_1 > \lambda_o$ along an orbit of X_{λ_o} :

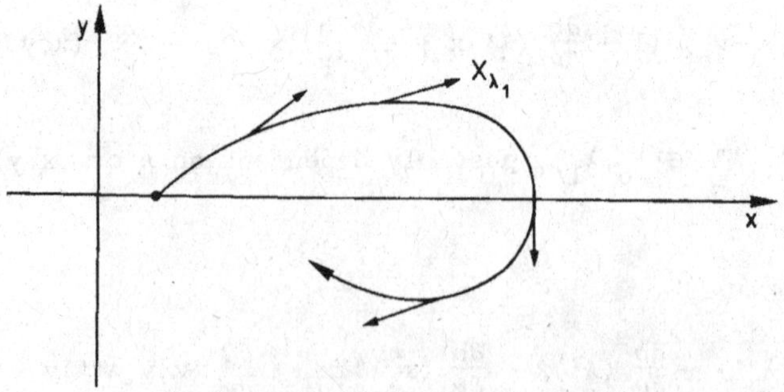

Figure 36

The following consequence can be formulated : take a fixed point $a \in Ox$ and a segment of orbit $\Gamma_\lambda = \bigcup_{t\in[0,T]} \phi_\lambda(t,a)$, $T > 0$ or $T < 0$, where $\phi_\lambda(t,a)$ is the flow of X_λ. Then, when ν increases , the arc Γ_λ turns in the positive sense with non zero speed :

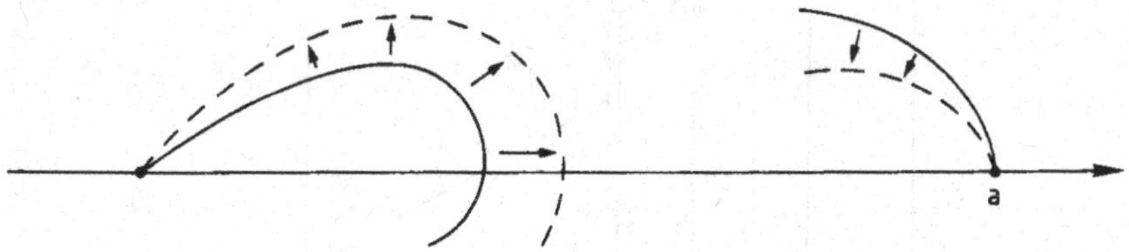

Figure 37

The same holds if a is a singular point for which Γ_λ is a separatrix.

This can be proved, using the following formula [A.L.], [S2] :

$$\frac{du}{d\nu}(\nu_o) = \frac{1}{\| X_{\lambda_o}(b)\|} \int_0^T (\exp [\int_0^t -\operatorname{div} X_{\nu_o}(\phi_{\lambda_o}(s,a))ds]) (\frac{\partial}{\partial \nu} < X^{\perp}_{\lambda_o}, X_\lambda > |_{\nu_o})dt$$

where $u(\nu)$ is the point of intersection near $b = \phi_{\lambda_o}(T,a)$ with σ of the segment of orbit Γ_λ through a and σ is a line segment through b tangent to and oriented by $X^{\perp}_{\lambda_o}(b)$ and parametrized by length.

2. Lines of loops and saddle-connections (SC,L)

Consider the saddle-case to begin with. We work with cubical neighborhoods of $0 \in \mathbb{R}^3$:

$$C_\epsilon = \{|\mu_1| \le \epsilon, |\mu_2| \le \epsilon \text{ and } |\nu| \le \epsilon_o\} \text{ where } \epsilon_o \text{ is fixed and } \epsilon < \epsilon_1 (\sim \epsilon_o^3)$$

goes to 0. The intersection of the SN-set with ∂C_ϵ is illustrated in Fig. 38.

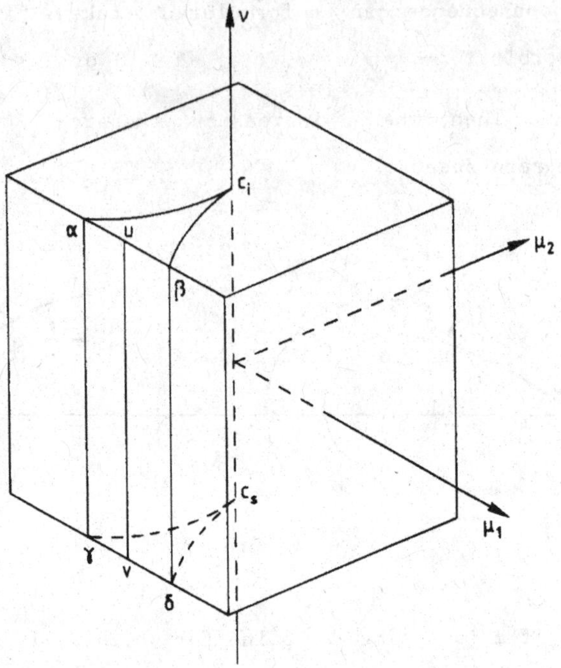

Figure 38

To describe the bifurcation set, it suffices to consider its intersection with the boundary of C_ϵ for ϵ small enough.
We know that c_i, $c_s \in \partial C_\epsilon$ are points of bifurcation of cuspidal type. For ϵ small enough the phase portrait of X_λ is known for λ on the horizontal faces $\nu = \pm \epsilon_0$ and in particular on $[\alpha,\beta]$ and $[\gamma,\delta]$. Let $[u,v]$ be a vertical segment from $u \in]\alpha,\beta[$ to $v \in]\gamma,\delta[$. The position of the separatrices of the two saddles s_1, s_2 are given in Figure 39 :

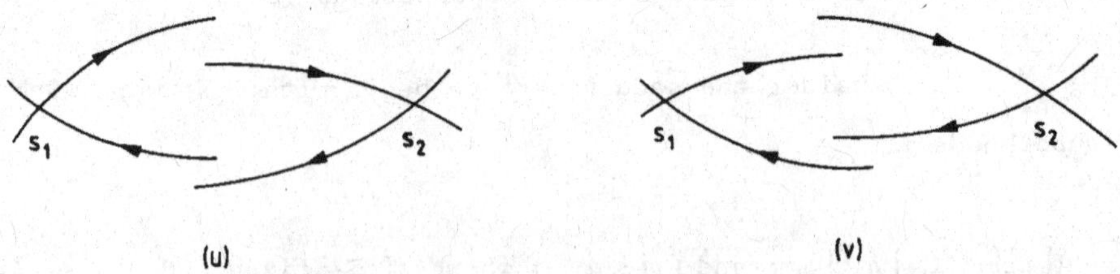

Figure 39

So, using the rotational property it follows obviously that when we go from u to v we have a generic bifurcation of superior and inferior saddle connections SC_s and SC_i. Parametrizing $]\alpha,\beta[$ by $\mu_1 \in]-\mu_1(\epsilon), \mu_1(\epsilon)[$ where $\mu_1(\epsilon) = (\frac{4}{27}\epsilon^3)^{1/2}$, we have lines of bifurcations on ∂C_ϵ given by graphs of functions $\mu_1 \to \nu_s(\mu_1,-\epsilon)$, $\nu_i(\mu_1,-\epsilon)$ $(\mu_2 = -\epsilon)$ and when $\mu_2 = -\epsilon$ goes to zero we have surfaces (called SC_s, SC_i) ,given by graphs $\nu_{s,i}(\mu_1,\mu_2)$. Actually, thanks to the conic structure to be established in the last paragraph, it suffices to look at one value of ϵ. Take some sufficiently small value of ϵ, and write $\nu_{s,i}(\mu_1,-\epsilon) = \nu_{s,i}(\mu_1)$. It is easy to see that the two lines SC_s, SC_i end at generic points : SNC = $SC(\pm\mu_1(\epsilon))$ of saddle-node connections. Also, by investigating the topological type near c_s and c_i, it is easy to see that at the left value $-\mu_1(\epsilon)$ we have $SC_s(-\mu_1(\epsilon)) > SC_i(-\mu_1(\epsilon))$ and at the right : $SC_s(\mu_1(\epsilon)) < SC_i(\mu_1(\epsilon))$. More can be said about the relative position of the lines SC_s and SC_i by using a "semi-rotational property" of the family X_λ with respect to the parameter μ_1. By "semi-rotational property" we mean that the typical situation of a rotational family will not occur in a full neighbourhood of the origin but only on a halfplane $\{y \geq 0\}$, resp. $\{y \leq 0\}$. Nevertheless this will suffice to give interesting information on the movement of the separatrices.

For that we take $\lambda'_0 = (\mu_1^0, \mu_2, \nu)$, $\lambda'_1 = (\mu_1^1, \mu_2, \nu)$ and calculate :

$$X_{\lambda'_1} - X_{\lambda'_0} = (\mu_1^1-\mu_1^0)(1 + \frac{\partial b}{\partial \mu_1}(\lambda')\, xy + x^3 y \frac{\partial h}{\partial \mu_1}(x,\lambda'') + y^2 \frac{\partial Q}{\partial \mu_1}(x,y,\lambda''')) \frac{\partial}{\partial y}$$

where $\lambda', \lambda'', \lambda''' \in [\lambda'_0, \lambda'_1]$, possibly dependent on x or (x,y)

Let $\phi'(x,y,\lambda',\lambda'',\lambda''') = \frac{\partial b}{\partial \mu_1}(\lambda')\, xy + x^3 y \frac{\partial h}{\partial \mu_1}(x,\lambda'') + y^2 \frac{\partial Q}{\partial \mu_1}(x,y,\lambda''')$.

We can choose the compact neighborhood $A \times B$ for the family X_λ such that $|\phi'(x,y,\lambda',\lambda'',\lambda''')| \leq \frac{1}{2}$ for $(x,y) \in A$ and $\lambda',\lambda'',\lambda''' \in B$, and hence

$$(\mu_1^1 - \mu_1^0)(1+\phi') \geq \frac{1}{2}(\mu_1^1 - \mu_1^0) \qquad \text{if } \mu_1^1 > \mu_1^0$$

With the same type of calculations as in the case of the ν-dependence, and taking into consideration that for the study of the movement of the separatrices one can work on the half-plane $\{y\geq 0\}$, resp. $\{y\leq 0\}$, it is possible to show that the lines SC_s and SC_i are also graphs with respect to the parameter ν.

By this the lines SC_s and SC_i will necessarily cross in exactly one point, which we denote by TSC (two saddles-cycle). Using the central rescaling in chapter VI we will locate and study this point of intersection. For the same reason there will be one point of intersection of SC_i and H. This point will also be studied in chapter VI, together with a point of intersection of H and SC_s. In Chapter VII we will conjecture that no other points of intersection occur between SC_s and H.

Now suppose that at some value μ_1 we have that : $SC_s(\mu_1) > SC_i(\mu_1)$. Then between these two values we find a unique loop bifurcation associated to the left saddle s_1, denoted by $L_\ell(\mu_1)$; see fig. 40. In the same manner, if $SC_s(\mu_1) < SC_i(\mu_1)$, we have a unique value $L_r(\mu_1)$ for a right loop connection. These connections are generic (transversal) by the rotational property in 1.

For μ_1 near $-\mu_1(\epsilon)$, $L_\ell(\mu_1)$ is the line of loops arriving at the Bogdanov-Takens point TB_ℓ; near $\mu_1(\epsilon)$, $L_r(\mu_1)$ is the line of loops ending at TB_r.

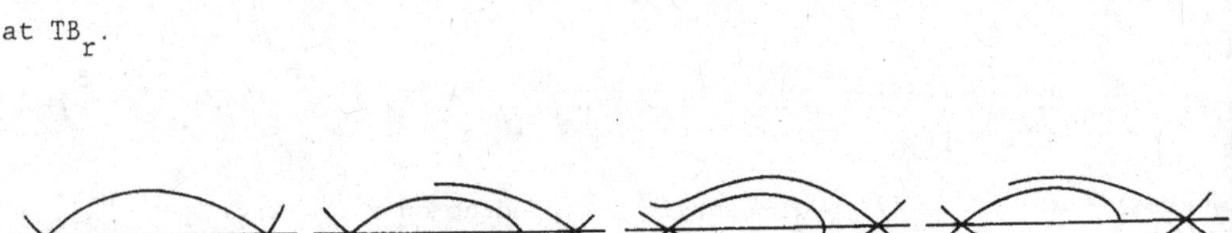

$SC_s(\mu_1)$ $L_\ell(\mu_1)$ $SC_i(\mu_1)$

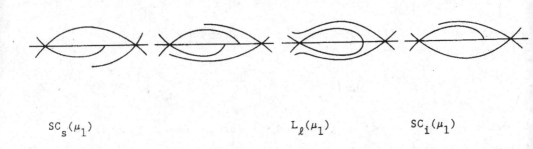

Fig. 40

In chapter VII we will conjecture that H and $L_\ell \cup L_r$ have no other points of intersection than the one between H and L_ℓ which will be studied in chapter VI using the central rescaling.

Fig. 41 illustrates the proposed bifurcation diagram for ϵ small enough, and one of the more complex possibilities which are discarded. See chapter VII.

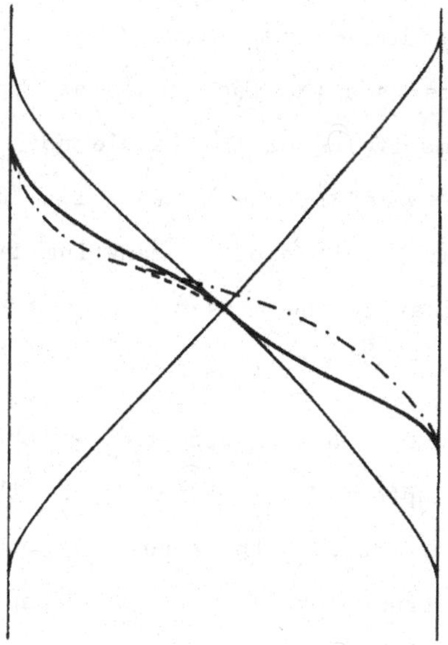

Proposed conjectural picture

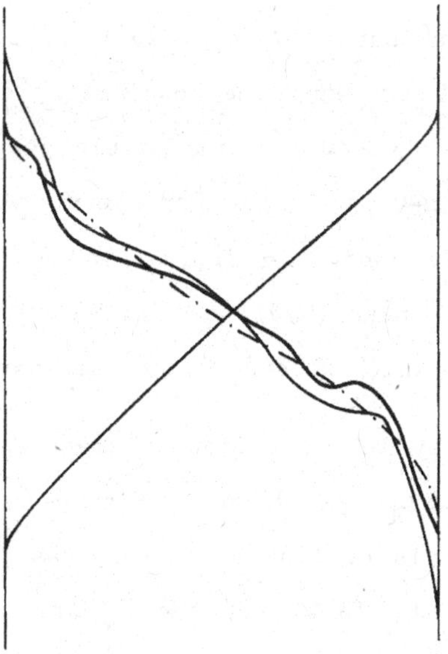

Discarded possibility (for H, L_ℓ, L_r, SC_1, SC_i)

Fig. 41

In the same way, in the focus case, we can easily prove the existence of lines of loop bifurcations L_r, L_ℓ (on ∂C_ϵ), using the structure of the family known in neighborhoods of c_i, c_s. The relative position of these two lines, as proposed in Figure 3, can be proven as follows : The divergence of X_λ is zero exactly along a regular curve cutting the x-axis transversally because of the implicit function theorem, using $b \neq 0$ (at least for small (x,y) and for small $\lambda = (\mu_1, \mu_2, \nu)$).

This prohibits the coexistence of a left- and a right loop at a same value of λ. L_ℓ as well as limit cycles surrounding only the left focus will occur when the line of zero divergence is to the left of the saddle, while

L_r and limit cycles surrounding only the right focus will occur when this line is to the right of the saddle. By this L_ℓ will be everywhere above L_r. This is a consequence of the fact that $b > 0$.

By the same rotational property with respect to ν it is now easy to prove that in between L_ℓ and L_r there exists a unique generic line L_i corresponding to loops selfconnecting the saddle s from below. (The factor $b > 0$ prevents the possibility of L_s). One uses the fact that the upper parts $(y \geq 0)$ of the stable and unstable manifolds of the saddle need to cut $\{x=0,\ y<0\}$ for parameter values "in between" L_ℓ and L_r (see fig. 3). Indeed, for $\lambda = (\mu_1,\mu_2,\nu)$ sufficiently small, all orbits emanating from points near (0,0) need to "spiral" before leaving the chosen neighbourhood A_o in which we study the phase portraits of X_λ. (see Chapter I.2.C).

Finally in the elliptic case there exist lines L_r and L_ℓ with L_r everywhere below L_ℓ, as shown in Figure 4; but in Chapter VII.D we will show that there is no line L_i in this case. The difference with the focus case comes from the fact that the X_λ are not longer transverse to ∂A_o (see Chapter I.2.C) and do not need to spiral before leaving ∂A_o, even if they start arbitrarily near (0,0). This has to do with the inequality $b > 2\sqrt{2}$.

3. Lines of boundary tangency

This paragraph concerns only the elliptic case. We choose a fixed neighborhood $A = \{x^2 + y^2 \leq \epsilon^2\}$, for a small $\epsilon > 0$. Recall that there are two fixed inner tangency points $(-\epsilon,0)$ and $(\epsilon,0)$ in ∂A. As mentioned above we know that the line L_r is everywhere below L_ℓ in the region I_ϵ of ∂C_ϵ (I_ϵ is the intersection of the sector I with ∂C_ϵ). Then, when ν increases, we must encounter generically unique lines of separatrix tangencies $ST_{r\ell}$, ST_r, ST_ℓ. See Fig. 42. Details will be given in Chapter VII.D. This has again to do with the b-factor, but the fact that, for the neighborhoods in consideration, exactly these lines occur and that they appear in the order mentioned, is a consequence of the sign chosen for ϵ_2 in $\epsilon_2 x^2 y \frac{\partial}{\partial y}$. It also explains the possibility of a line $DT = DT_s$ and the impossibility of a line DT_i (i for inferior).

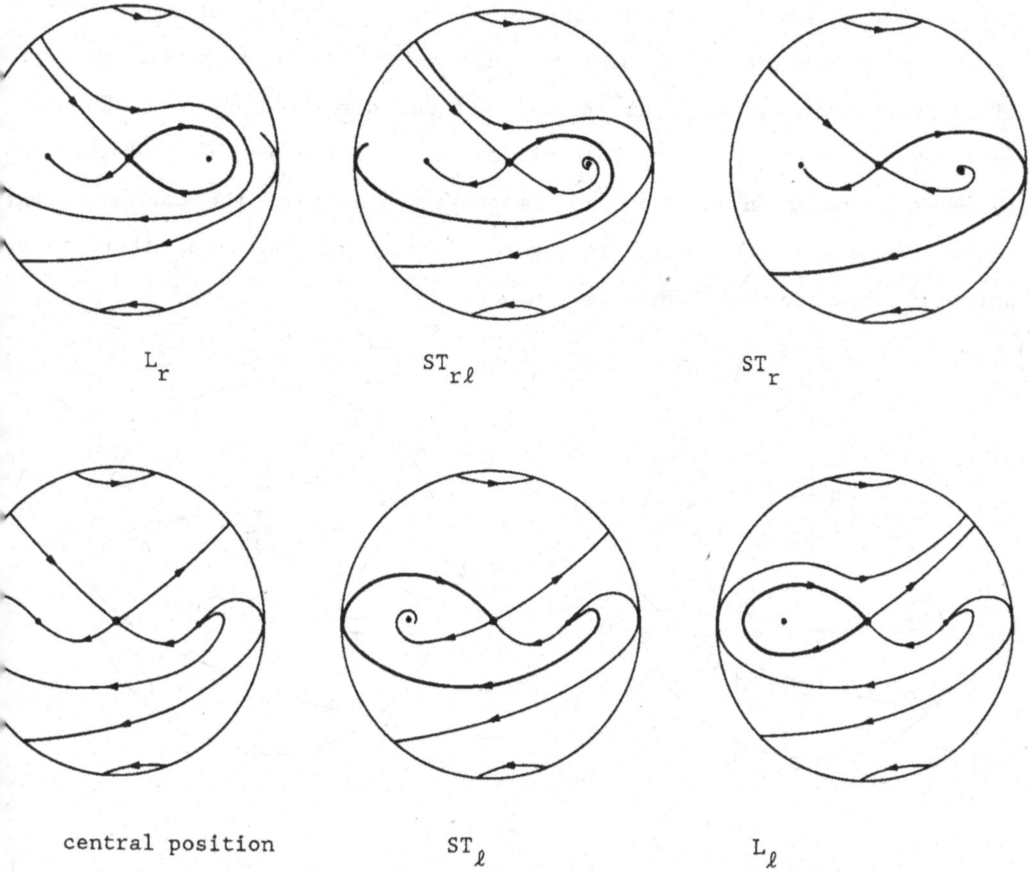

Fig. 42

The sequence of pictures in Figure 42 is easily established, taking into account the knowledge near c_i, c_s and the fact that the orbit starting at β (the right tangency point) cuts transversaly ∂A below α (the left tangency point).

Then, it is easy to show that the end points of these bifurcation lines L_r, $ST_{r\ell}$, ST_r, ST_ℓ, L_r are generic codimension 2 bifurcation points. On the right side the line ST_ℓ ends at the point $DCST_c$ where the two lines CT_ℓ, DT_s in the region E_ϵ $(E \cap \partial C_\epsilon)$ begin. On the left side the lines $ST_{r\ell}$ and ST_r are also changed, when crossing SN_ℓ, into the two lines CT_ℓ, DT_s entering the region E_ϵ.

Now it can be proved that the pair of lines CT_{ℓ} and DT_s starting to the right and to the left of I_{ϵ} into E_{ϵ} connect in E_{ϵ} to form a unique pair of lines, everywhere transversal to the vertical direction on ∂C_{ϵ}. To show this take any such transversal [u,v] (parallel to the axis $O\nu$), whith u on $\{\nu = -\epsilon_0\}$ and v on $\{\nu = \epsilon_0\}$. The phase portrait of the corresponding vector fields are illustrated in Figure 43. Recall that in E_{ϵ} there is a unique critical point, a focus or a node.

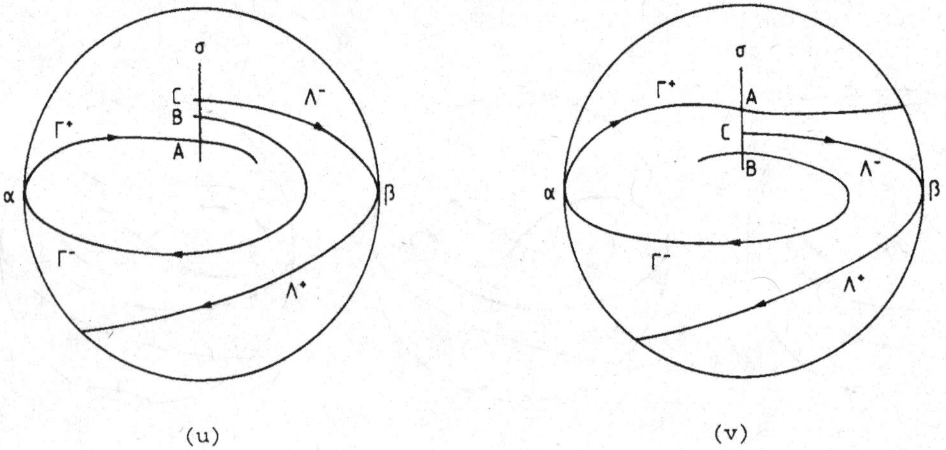

Figure 43

In the 2 phase portraits, Λ^+ keeps the same position as shown in the 2 pictures; we see that the intersection point A of the right trajectory Γ^+ from α with a vertical segment σ moves upwards, from below and ending above the points B and C, intersection with σ of the trajectories Γ^- and Λ^- through α and β respectively. When A passes through B, we have a limit cycle tangent to the boundary. When A passes through C, we have a double tangency. This second bifurcation is the generic line DT_s (s for superior). We do however not know for which parameter values the limit cycle is actually hyperbolic at the moment of tangency with the boundary. We conjecture that this is everywhere the case except for one value where we have a double limit cycle and where the unfolding is like we described it in Chapter IV, under the name "Double Cycle Tangency" (DCT).

The rest of the picture in Figure 4, concerning the lines CT_ℓ, DT_s (relative position with the line H, existence of the line DC as a unique regular line,...) is also conjectural.

4. The set of limit cycles

Consider the saddle case and assume the notations of paragraph 2. Take a value $\mu_1 \in [-\mu_1(\epsilon), \mu_1(\epsilon)]$ where $SC_s(\mu_1) \geq SC_i(\mu_1)$. Take any $x \in [s_1(\mu_1), e(\mu_1)]$ ($s_1(\mu_1)$ is the left saddle and $e(\mu_1)$ the intermediate point). If we look at the limiting position of ν, we see that the first return point $P_\lambda(x)$ of the trajectory through x on [s,e] is first to the right and next to the left of x when we let ν increase. So, for a unique value of ν there exists a limit cycle passing through x. We obtain a C^∞ map $\nu(x,\mu_1)$: $x \in [s_1(\mu_1), e(\mu_1)] \to \mathbb{R}$ such that a limit cycle of $X_{(\mu_1,\nu)}$ passes through x. This map is defined and C^∞ in (x,μ_1) for each μ_1 such that $SC_s(\mu_1) \geq SC_i(\mu_1)$.

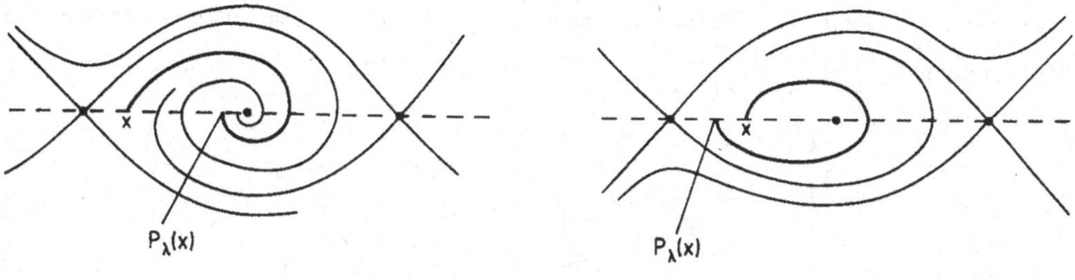

Fig. 44

If $SC_s(\mu_1) \leq SC_i(\mu_1)$ we may define $\nu(x,\mu_1)$ for $x \in [e(\mu_1), s_2(\mu_1)]$. If $SC_s(\mu_1) = SC_i(\mu_1)$ we may define $\nu(x,\mu_1)$ taking any side of $e(\mu_1)$. Coming back to the first case, let us draw the graph of $x \to \nu(x,\mu_1)$. The end points correspond to the Hopf bifurcation for $x = e(\mu_1)$ and to a left loop for $x = s_1(\mu_1)$. For any ν, the counter image $\nu^{-1}(\nu)$ is formed by the values x where $X_{(\mu_1,\nu)}$ has a limit cycle.

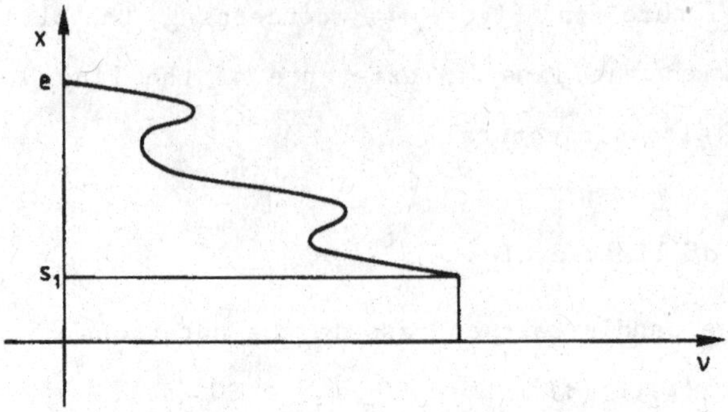

Fig. 45

The union of all these graphs (with the obvious identification when $SC_s(\mu_1) = SC_i(\mu_1)$) is a surface $\Sigma \subset \mathbb{R}^3$ (space of (x, μ_1, ν)) diffeomorphic to a disk. (We let $\mu_2 = -\epsilon$ fixed to simplify the description). The projection of Σ on the parameter space (μ_1,ν) covers the set of parameter values for which there exist limit cycles. The boundary projects on $L_\ell \cup H \cup L_r$ and the critical locus of this projection is the set of (μ_1,ν) where there exist non-hyperbolic limit cycles. We will conjecture in Chapter VII that the projection has only one fold which creates a triangular region in the (μ_1,ν)-space in which one finds two limit cycles.

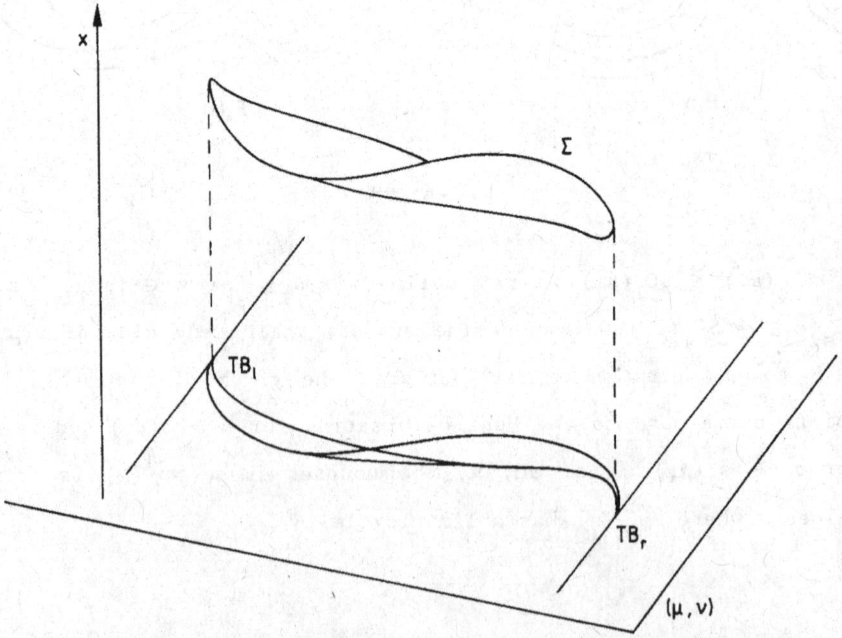

Fig. 46

Analogous things may be said for the focus and elliptic cases. The conjectures concerning the projection of Σ are easily formulated after looking at Figures 3-4. **In each case, Σ is diffeomorphic to a disk.**

V.E. The principal rescaling

This rescaling, indicated in the introduction, is defined by the formula : $x = t\bar{x}$, $y = t^2\bar{y}$; $\mu_2 = t^2\bar{\mu}_2$, $\mu_1 = t^3\bar{\mu}_1$, $\nu = t\bar{\nu}$. For each $t > 0$, this defines a change of coordinates and in the new coordinates $(\bar{x},\bar{y})$ with the new parameter $\bar{\lambda} = (\bar{\mu}_1, \bar{\mu}_2, \bar{\nu})$, we have :

$$\frac{1}{t} X_{(\bar{\lambda},t)} = \bar{X}^P_{\bar{\lambda}} + O(t) \tag{55}$$

$$\text{where } \bar{X}^P_{\bar{\lambda}} = \bar{y}\frac{\partial}{\partial\bar{x}} + (\epsilon\,\bar{x}^3 + \bar{\mu}_2\bar{x} + \bar{\mu}_1 + \bar{y}\,(\bar{\nu} + b\bar{x}))\frac{\partial}{\partial\bar{y}} \tag{56}$$

and $O(t)$ is a family of vector fields of order t.

Notice that in the preceeding paragraphs the calculations do not use the term $x^2y\frac{\partial}{\partial y}$, which is not present in $\bar{X}^P_{\bar{\lambda}}$. Actually (and below we are going to briefly discuss this point) it is possible to recover for the family $\bar{X}^P_{\bar{\lambda}}$ all the bifurcations found above, except perhaps for the boundary phenomena, which depend on the choice of the neighborhood and will not be considered in this framework.

Notice first that to study a neighborhood in the original parameter it suffices to take $\bar{\lambda} \in \bar{S} = \{\bar{\mu}_1^2 + \bar{\mu}_2^2 + \bar{\nu}^2 = 1\}$ or $\bar{\lambda} \in \partial C$, where C is a cubical neighborhood as in D, and choose t small enough. We limit ourselves to the bifurcation appearing in the plane $\bar{\mu}_2 = -1$ and $(\bar{\mu}_1,\bar{\nu}) \in K$, some compact in the plane $(\bar{\mu}_1,\bar{\nu})$. This gives the essential of the bifurcation set in the saddle case ; in the other cases we must take also $\bar{\mu}_2 = 1$ but the study is very similar and will be omitted. So, take $\epsilon = 1$ in what follows.

The line T (on which the trace is zero at same singular point) is given by $\bar{x}_o^3 - \bar{x}_o + \bar{\mu}_1 = 0$ and $\bar{\nu} + b\bar{x}_o = 0$. The line SN of saddle-node is $\bar{\mu}_1 = \pm\frac{2}{3\sqrt{3}}$ and for this value, the degenerate singular point is $(\pm\frac{1}{\sqrt{3}},0)$. The Bogdanov-Takens TB_ℓ for example is in $\bar{\mu}_1 = -\frac{2}{3\sqrt{3}}$, $\bar{\nu} = \frac{b}{\sqrt{3}}$.

Taking coordinates (X,Y) around the degenerate point $(-\frac{1}{\sqrt{3}},0)$, given by $\bar{x} = X - \frac{1}{\sqrt{3}}$, $\bar{y} = Y$, and local parameters (M,N) given by $\bar{\mu}_1 = -\frac{2}{3\sqrt{3}} + M$, $\bar{\nu} = \frac{b}{\sqrt{3}} + N$, we have that

$$\bar{X}^P_{\bar{\lambda}} = Y \frac{\partial}{\partial X} + [(M - \sqrt{3} X^2 + X^3] + Y(N + bX) \frac{\partial}{\partial Y} \tag{57}$$

This family is already in the normal form given in IV.B.3. Compare with the proof given above in C.

It is also very easy to verify the rotational property of $\bar{X}^P_{\bar{\lambda}}$ with respect to $\bar{\nu}$: taking $\bar{\lambda}_1 = (\bar{\mu}_1, \bar{\nu}_1)$ and $\bar{\lambda}_0 = (\bar{\mu}_1, \bar{\nu}_0)$ we have :

$$\bar{X}_{\bar{\lambda}_1} - \bar{X}_{\bar{\lambda}_0} = (\bar{\nu}_1 - \bar{\nu}_0)\, \bar{y} \frac{\partial}{\partial \bar{y}} \quad \text{and}$$

$$\frac{\partial}{\partial \bar{\nu}} < X^{\perp}_{\bar{\lambda}_0} , X_{\bar{\lambda}} > = \bar{y}^2 \tag{58}$$

So, all the results established in paragraph D concerning the lines of saddle-connection and loops, are also true for the family $\bar{X}^P_{\bar{\lambda}}$. The same will also hold for the boundary tangencies in well chosen neighborhoods.

Since these lines as well as their extremal points are generic (i.e. transversal) bifurcations, a simple implicit function argument allows to conclude : to each generic bifurcation line $\bar{\sigma}$ of the family $\bar{X}^P_{\bar{\lambda}}$ corresponds a bifurcation surface diffeomorphic to $\bar{\sigma} \times [0,\epsilon] \subset \mathbb{R}^2 \times [0,\epsilon]$ (in the parameter space $(\bar{\mu}_1, \bar{\nu}, t)$) for $\epsilon > 0$ small enough. There also corresponds a bifurcation surface of conic shape in the parameter space (μ_1,μ_2,ν) given by :

$$\{(t^3 \bar{\mu}_1, -t^2, t\,\bar{\nu}) \mid (\bar{\mu}_1, \bar{\nu}, t) \in \sigma \times [0,\epsilon]\} \tag{59}$$

Notice that the whole bifurcation set cannot be studied in this principal rescaling. In fact, the vector field $\bar{X}^P_{\bar{\lambda}}$, along the line $(\bar{\mu}_1 = \bar{\nu} = 0)$, due to the absence of the term $\bar{y}\,\bar{x}^2 \frac{\partial}{\partial \bar{y}}$, admits the symmetry $(\bar{x},\bar{y}) \to (-\bar{x},\bar{y})$.

Along this line, we have to reintroduce this term. This will be done by means of the central rescaling studied in the following chapter.

CHAPTER VI : THE CENTRAL RESCALING

VI.A. Definition and basic properties

As mentioned in the introduction this rescaling is given by :
$x = \tau x'$, $y = \tau^2 y'$, $\mu_1 = \tau^4 \mu_1'$, $\mu_2 = \tau^2 \mu_2'$, $\nu = \tau^2 \nu'$.
The family X_λ written in normal form is given by formula (0) in Chapter V :

$$\frac{1}{\tau} X_\lambda = y' \frac{\partial}{\partial x'} + ((\epsilon x'^3 + \mu_2' x' + bx'y') + \tau(\mu_1' + \nu' y' + y'x'^2) + y' O(\tau^2))\frac{\partial}{\partial y'} \qquad (1)$$

For $\tau > 0$, X_λ is C^∞ equivalent to the family $\frac{1}{\tau} X_\lambda$ written above.
This family for $\tau \to 0$ tends to

$$X^S = y' \frac{\partial}{\partial x'} + (\epsilon x'^3 + \mu_2' x' + bx'y') \frac{\partial}{\partial y'} \qquad (2)$$

This rescaling will be useful when X^S has non degenerate critical points, i.e. : $\mu_2' \neq 0$. To study such a situation it suffices to suppose that $\mu_2' = \pm 1$, $(\mu_1', \nu') \in K$, an arbitrarily large compact subset in $\mathbb{R}^2$, and $\tau \in]0,T]$, T small enough but positive.
Recall that the central rescaling may be seen as a blowing up in the coordinates of the principal rescaling, given by $t = \tau$, $\bar{\mu}_1 = \tau\mu_1'$, $\bar{\nu} = \tau\nu'$.
The region $C^S = \{(\mu_1', \nu', \tau) \in K \times [0,T]\}$ corresponds in the parameter of the principal rescaling to :

$$\{(t\mu_1', t\nu', t) \mid (\mu_1', \nu') \in K, \ t \in]0,T]\} \qquad (3)$$

This cone cuts the planes $\mathbb{R}^2 \times \{t\}$ along the compact set $tK \times \{t\}$ whose diameter goes to zero with t.
In what follows we omit the primes for x',y'.

$$\text{Write } X^S = y\frac{\partial}{\partial x} + (\epsilon x^3 \pm x + bxy) \frac{\partial}{\partial y} \qquad (4)$$

This vector field is not Hamiltonian since $\operatorname{div} X^S(x,y) = bx \neq 0$. But Oy is a global symmetry axis for its phase portrait. This is related to the following more precise property. If we consider the dual form $\omega_S = ydy - (\epsilon x^3 \pm x + bxy)dx$, we have that :

$$\omega_S = U_*(\Omega_S) \text{ where } \Omega_S = ydy - (2\epsilon X \pm 1 + by)\, dX \tag{5}$$

$$\text{and } U(x,y) = (\frac{x^2}{2}, y) \tag{6}$$

Later on, we will use this formula to obtain analytic first integrals for X^S. It means that ω_S is the pullback of the linear form Ω_S by the fold map (6). The phase portrait follows immediately from the position of the fold-line $\{x = 0\}$ with respect to the singular point of Ω_S. In Figures 47 and 48 we represent the different possibilities.

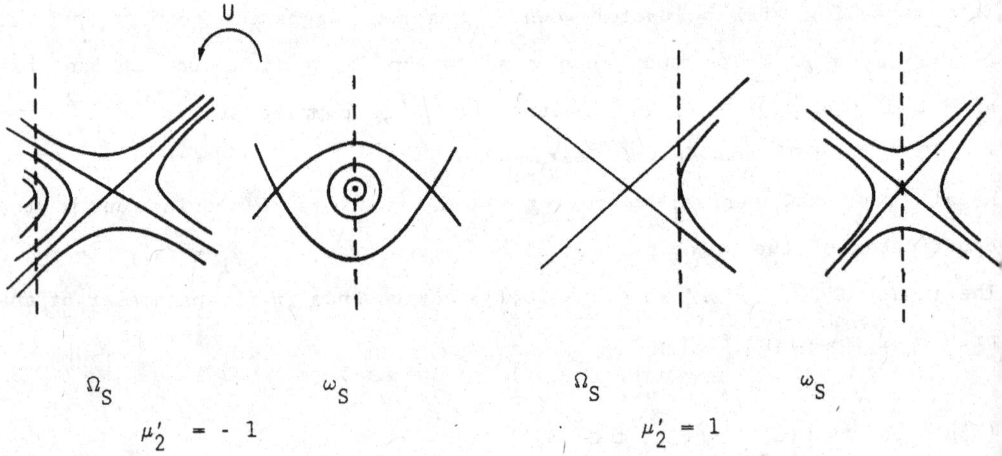

Saddle case

Fig. 47

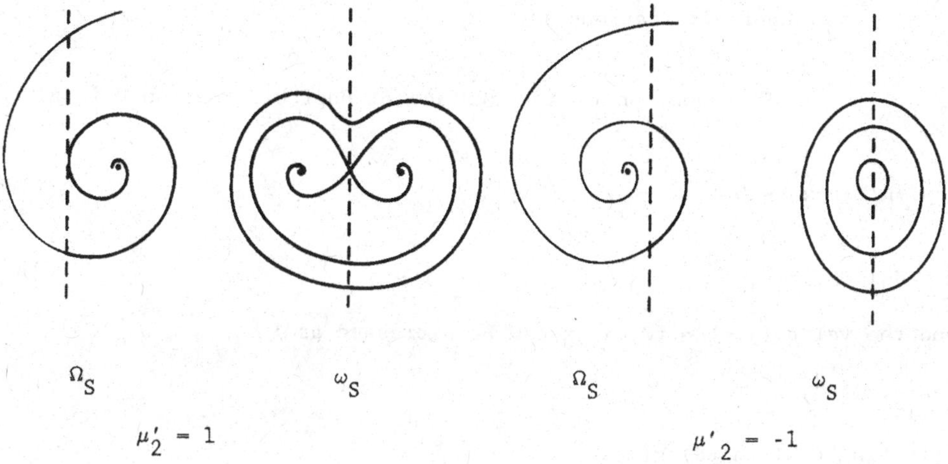

Focus case

Fig. 48

Remark : The picture in the elliptic case is similar to Fig. 48 with foci replaced by nodes.

VI.B. <u>The saddle case</u>

Here we have $\epsilon = 1$. The case $\mu_2' = 1$ is of no interest for the bifurcation analysis because X^S is structurally stable. So we consider only $\mu_2' = -1$:

$$\begin{cases} X^S = y \dfrac{\partial}{\partial x} + (x^3 - x + bxy) \dfrac{\partial}{\partial y} \\ X_{(\mu_1', \nu', \tau)} = X^S + (\tau(\mu_1' + \nu' y + x^2 y) + O(\tau^2) y) \dfrac{\partial}{\partial y} \end{cases} \tag{7}$$

Recall that $b > 0$.

1. Hopf bifurcations

The equation of the surface of Hopf bifurcation H is given by :

$$\begin{cases} bx + \tau\nu' + \tau x^2 = 0 & (8) \\ x^3 - x + \tau\mu_1' = 0 & (9) \end{cases}$$

Around the value $(x,\tau) = (0,0)$, x can be expressed as :

$$x = \tau\mu_1' + O(\tau^2) \tag{10}$$

Substituting (10) in (8) gives :

$$\tau(b\mu_1' + \nu' + O(\tau)) = 0 \tag{11}$$

Hence, the limit of the surface H for $\tau \to 0$ is the line (called also H) :

$$b\mu_1' + \nu' = 0 \tag{12}$$

A candidate for the Hopf bifurcation of codimension 2 was found in V.B (formulas (28)). Making $t = \frac{\tau}{(3b)^{1/2}}$ in these formulas, one gets :

$$\begin{cases} \nu = \frac{1}{3}\tau^2 + O(\tau^4) \\ \mu_1 = -\frac{1}{3b}\tau^4 + O(\tau^6) \\ \mu_2 = -\tau^2 + O(\tau^4) \end{cases} \tag{13}$$

As $\mu_2' = -1$, we obtain, using the rescaling formulas that :

$$\nu' = \frac{1}{3} + O(\tau) \quad , \mu_1' = -\frac{1}{3b} + O(\tau) \tag{14}$$

which gives the point $DH = (\frac{1}{3}, -\frac{1}{3b})$ as limit position for $\tau \to 0$.
We show in the next paragraph that this bifurcation point is generic.

2. Integrating factor and Abelian integral

This paragraph relies on the work of Zoladek [Z1]. First notice that the linear form Ω_S may be written as follows :

$$\Omega_S = y\,dy - \frac{1}{2}(v + by)\,dv$$

$$\text{where } v = x^2 - 1 \tag{15}$$

The dual vector field of Ω_S , in the coordinates (v,y), has the following orbit equations :

$$\begin{cases} \dot{v} = y \\ \dot{y} = \frac{1}{2}(v+by) \end{cases} \tag{16}$$

The origin is a hyperbolic saddle with eigenvalues : $\frac{1}{4}(b \pm (b^2+8)^{1/2})$ and eigenvectors $(1, \frac{1}{4}(b \pm (b^2+8)^{1/2})$. Write :

$$\begin{cases} \bar{\alpha} = \frac{1}{4}(b + (b^2+8)^{1/2}) & (\bar{\alpha} > 0) \\ \bar{\beta} = \frac{1}{4}(b - (b^2+8)^{1/2}) & (\bar{\beta} < 0 \text{ and } 0 < |\bar{\beta}| < \bar{\alpha}) \\ V = y - \bar{\alpha}v & \\ Y = \bar{\beta}v - y & \end{cases} \tag{17}$$

Notice that V, Y are diagonalizing coordinates for the vector field. It is easy to verify that :

$$(\bar{\beta} - \bar{\alpha})\,\omega_S = \bar{\alpha}\,Y\,dV - \bar{\beta}\,V\,dY \tag{18}$$

and then :

$$\frac{V^{\bar{\alpha}}}{Y^{\bar{\beta}}}\left(\frac{\bar{\beta}-\bar{\alpha}}{YV}\right)\omega_S = \frac{V^{\bar{\alpha}}}{Y^{\bar{\beta}}}\,\frac{1}{YV}\,(\bar{\alpha}\,YdV - \bar{\beta}VdY)$$

$$= \bar{\alpha}\,V^{\bar{\alpha}-1}\,Y^{-\bar{\beta}}\,dV - \bar{\beta}\,Y^{-\bar{\beta}-1}\,V^{\bar{\alpha}}\,dY = d\left(\frac{V^{\bar{\alpha}}}{Y^{\bar{\beta}}}\right) \tag{19}$$

So we can use :

$V^{\bar{\alpha}-1}\,Y^{-\bar{\beta}-1}$ as an integrating factor and $\frac{1}{\bar{\beta}-\bar{\alpha}}\,V^{\bar{\alpha}}\,Y^{-\bar{\beta}}$ as a Hamiltonian for the vector field X^S, with $V = y - \bar{\alpha}(x^2-1)$ and $Y = \bar{\beta}(x^2-1) - y$ (20)

We can also work with

$$\begin{cases} K = V^{\bar{\bar{\alpha}}-1}\, Y^{\bar{\bar{\beta}}-1} \text{ as an integrating factor and} \\ H = -\dfrac{1}{\bar{\bar{\alpha}}+\bar{\bar{\beta}}}\, V^{\bar{\bar{\alpha}}} Y^{\bar{\bar{\beta}}} \text{ as Hamiltonian} \end{cases} \tag{21}$$

where $\bar{\bar{\alpha}} = r\bar{\alpha}$, $\bar{\bar{\beta}} = -r\bar{\beta}$ for any $r \neq 0$

Remark : We are endebted to A. Lins Neto for suggesting the above Hamiltonians, also found independently by Zoladek [Z1]. Notice that the eigenvalues of X^S are $\lambda_1 = -\bar{\beta}$, $-\xi_1$ with $\xi_1 = \bar{\alpha}$ at the saddle s_1 (-1,0) and $\lambda_2 = \bar{\alpha}$, $-\xi_2 = -\bar{\beta}$ at the saddle s_2 (1,0). Two common separatrices $\gamma_s = \{V = 0\}$ and $\gamma_i = \{Y = 0\}$ form the double saddle singular cycle γ. We may observe that the assymmetry between the 2 branches γ_i, γ_s increases with b.

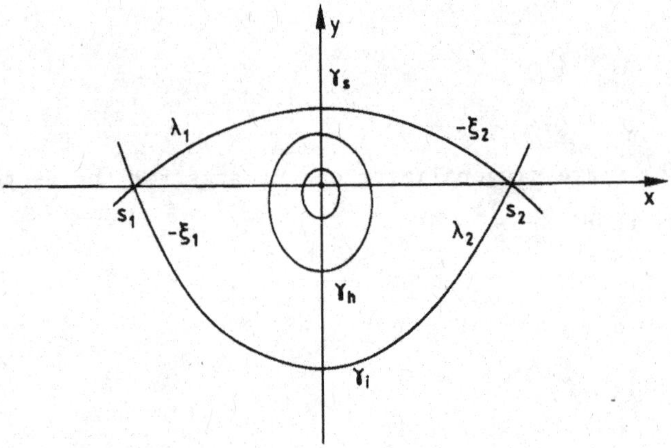

Fig. 49

Return now to the family in dual form :

$$\omega_{(\mu_1',\nu',\tau)} = \omega_S - \tau\,(\mu_1' + \nu' y + x^2 y)\,dx + O(\tau^2)ydx \tag{22}$$

with $\omega_S = ydy - (x^3 - x + bxy)dx$

Any integrating factor $K = V^{\bar{\bar{\alpha}}-1}\, Y^{\bar{\bar{\beta}}-1}$ is non zero in the interior of the disk D bounded by the singular cycle γ. So, taking any compact set $B' \subset \text{int } D$, we may replace $\omega_{(\mu_1',\nu',\tau)}$ by $K\omega_{(\mu_1',\nu',\tau)}$ to study the cycles near B'.

$$K\omega_{(\mu_1',\nu',\tau)} = dH - \tau K(\mu_1' + \nu' y + x^2 y)dx + O(\tau^2)Kydx \qquad (23)$$

with $H = -\frac{1}{\bar{\bar{\alpha}}+\bar{\bar{\beta}}} V^{\bar{\bar{\alpha}}} Y^{\bar{\bar{\beta}}}$

The family of forms (23) is of the perturbed Hamiltonian type presented in the introduction of Chapter IV. The value of H increases from $H(0,0) = -\frac{1}{\bar{\bar{\alpha}}+\bar{\bar{\beta}}} V(0,0)^{\bar{\bar{\alpha}}} Y(0,0)^{\bar{\bar{\beta}}} = h_o < 0$ to the value 0 on γ. We parametrize a segment σ from $e = (0,0)$ to a point on γ, and transversal to H ($\sigma = [e,s_2]$ for example) by the values of H : $\sigma \simeq [h_o,0]$. Let γ_h be the cycle of H through the point $h \in \sigma$ and let :

$$\int_{\gamma_h} \omega_D = J(h,\lambda') = \mu_1' J_o + \nu' J_1 + J_2 \quad ; \quad \lambda' = (\mu_1', \nu') \qquad (24)$$

where $\omega_D = K(\mu_1' + \nu' y + x^2 y)\, dx$

Let also any $h_1 \in \,]\, h_o,0\, [$ (One may suppose h_1 near 0) and take $B' = D_{h_1} = \{H \le h_1\} \subset \operatorname{int} D$.

We recall that the study of limit cycles through $[h_o,h_1]$ is based on the perturbation lemma (see the Introduction of Chapter IV), which states that if $P_t(h)$ is the Poincaré-map of $X_{(\mu_1',\nu',t)}$ on $[h_o,h_1]$, then

$$\frac{P_t(h) - h}{t} = J + O(t) \qquad (25)$$

This asymptotic formula allows the reduction of the study of fixed points of $P_t(h)$ to the study of zeroes of the Abelian integral J.

Let $\frac{J}{J_o} = \mu_1' - P_1\nu' - P_2 \quad (P_1 = -\frac{J_1}{J_o},\ P_2 = -\frac{J_2}{J_o}) \qquad (26)$

Because $J_o \sim (h - h_o)$ and $J_1,J_2 = O(h - h_o)$, the ratio is also an analytic function and may be used to locate the bifurcations.

The limit Hopf line H is given by :

$$H : \frac{J}{J_o}(h_o) = \mu_1' - P_1(h_o)\nu' - P_2(h_o) = 0 \qquad (27)$$

It is easily proved that $P_1(h_o) = -\frac{1}{b}$ and $P_2(h_o) = 0$. So we recover equation (12) for H.

To obtain more information, we bring our system to the form studied in [Z1] :

$$\begin{aligned} \dot{x} &= y \\ \dot{y} &= -x + \alpha x^3 + xy + \beta_o + \beta_1 y + \beta_2 x^2 y \end{aligned} \tag{28}$$

It suffices to put $b = \frac{1}{\sqrt{\alpha}}$, $\beta_o = \frac{1}{\sqrt{\alpha}} \tau \mu_1'$, $\beta_1 = \tau\nu'$, $\beta_2 = \alpha\tau$, (29)
and make the transformation $(x,y) \to (\frac{x}{\sqrt{\alpha}}, \frac{y}{\sqrt{\alpha}})$.

The choice of Hamiltonian in [Z1] corresponds to $r = \frac{2}{b}$.
Finally Zoladek used another transformation in coordinates and parameters (called now ϵ_o, ϵ_1, ϵ_2) which replaces the problem for J by an equivalent problem for $I = \epsilon_o I_o + \epsilon_1 I_1 + \epsilon_2 I_2$: one passes from $(\beta_o, \beta_1, \beta_2)$ to $(\epsilon_o, \epsilon_1, \epsilon_2)$ and from (J_o, J_1, J_2) to (I_o, I_1, I_2) by linear isomorphisms ; the Hamiltonian parameter, called c, runs over [0,1], where 1 is the value at the center.

In [Z1], it was proved that the curve $c \in]0,1] \overset{Q}{\to} (Q_1(c), Q_2(c))$ where $Q_1 = -\frac{I_1}{I_o}$, $Q_2 = -\frac{I_2}{I_o}$, is simple and strictly convex on]0,1]. This property is invariant under linear isomorphism. So the map
$h \in [h_o, 0[\overset{P}{\to} (P_1(h), P_2(h))$ is also strictly convex. Also it was proved in [Z1] that Q_1 is strictly monotonic. The same is true for P_1.

These properties have been used in [DRS] to prove a result similar to the following one. The arguments will not be repeated here.

- The fact that P is strictly convex at $h = h_o$ implies that the point DH is a generic codimension 2 Hopf bifurcation. The point DH is defined by $\mu'_1 - \nu' P_1(h_o) - P_2(h_o) = 0$ and $\nu' = -\frac{P_2'}{P_1'}(h_o)$.
- The envelope curve of the h-parameter family of lines

$\delta_h = \{\mu_1' - \nu' P_1(h) - P_2(h) = 0\}$ which is also strictly convex, is the limit for $\tau \to 0$ of a generic bifurcation line of semi-stable cycles : DC.

- Let h_1 be a fixed value $h_o < h_1 < 0$ and let $T(h_1)$ be the triangular open domain limited by $H = \delta_{h_o}$, δ_{h_1} and DC. For τ sufficiently small, we obtain, using the Implicit Function Theorem a nearby deformed domain T_τ such that for each $\lambda' \in T_\tau$, the vector field $X_{\lambda',\tau}$ has exactly 2 limit cycles.

Remark : We cannot apply directly the preceeding result to $h_1 = 0$, because the Hamiltonian H is not analytic along the singular cycle γ. We consider this limit situation in paragraph 5. The limit position of δ_h, for $h \to 0$, reduces to $\int_\gamma \omega_D = \int_{\gamma_s} \omega_D = 0$ (Because the integrating factor used above is identically zero in γ_i). We will see in the next paragraph that this last equation is the limit for $\tau \to 0$ of the equation for the line L_ℓ of left loops, for the line L_r of right loops and also the limit of the line SC_s of superior saddle connections).

The results of this paragraph are summarized in Figure 50 : (the limit position of the points TSC and SC_s will be calculated in paragraphs 4 and 3, respectively).

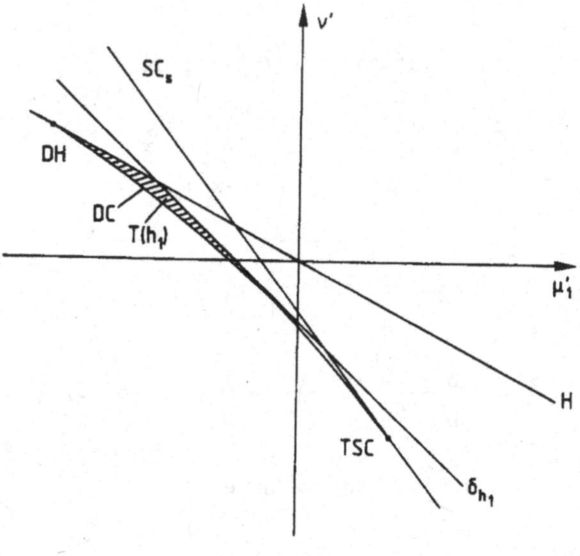

Figure 50

3. Saddle connections (SC_s, SC_i)

3.1. The line SC_s (superior saddle connection)

As it was recalled in Chapter IV, the equation of the superior saddle connections, at the limit $\tau \to 0$, is given by :

$$S(\lambda') = \int_{\gamma_s} \omega_D = 0 \tag{30}$$

which is :

$$\int_{-1}^{1} (y-\bar{\alpha}(x^2-1))^{\bar{\bar{\alpha}}-1} (\bar{\beta}(x^2-1)-y)^{\bar{\bar{\beta}}-1} (\mu_1' + \nu' y + x^2 y)\, dx \tag{31}$$

To make the calculation meaningful, the Hamiltonian must be non singular at any point of int γ_S. So, we need to choose :

$$\bar{\bar{\beta}} = 1, \text{ i.e } r = -\frac{1}{\bar{\beta}} \quad \text{(see 21)} \tag{32}$$

Since $y = \bar{\beta}(x^2-1)$, we have to calculate :

$$\int_{-1}^{1} [(\bar{\beta}-\bar{\alpha})\ (x^2-1)]^{-(\frac{\bar{\alpha}}{\bar{\beta}}+1)} (\mu_1'+\nu'\bar{\beta}(x^2-1) + \bar{\beta}x^2(x^2-1))\, dx =$$

$$= (\bar{\alpha}-\bar{\beta})^{-(\frac{\bar{\alpha}}{\bar{\beta}}+1)} [(\int_{-1}^{1} (1-x^2)^{-\frac{\bar{\alpha}}{\bar{\beta}}-1} dx)\ \mu_1' - (\bar{\beta}\int_{-1}^{1} (1-x^2)^{-\frac{\bar{\alpha}}{\bar{\beta}}} dx)\ \nu' -$$

$$(\bar{\beta}\int_{-1}^{1} x^2(1-x^2)^{-\frac{\bar{\alpha}}{\bar{\beta}}} dx)] \tag{33}$$

Let us denote by $c = -\frac{\bar{\alpha}}{\bar{\beta}} = -\frac{b + (b^2+8)^{1/2}}{b - (b^2+8)^{1/2}} = \frac{1}{8}(b + (b^2+8)^{1/2})^2 > 1$ (34)

Recall that $\bar{\beta} = \frac{1}{4}(b - (b^2+8)^{1/2})$. So, we have

$$S(\lambda') = A'\mu'_1 + B'\nu' + C' \tag{35}$$

with $A' = \int_0^1 (1-x^2)^{c-1}\,dx$, $\quad B' = -\bar{\beta}\int_0^1 (1-x^2)^c dx$

and $C' = -\bar{\beta}\int_0^1 x^2(1-x^2)^c dx = -\frac{\bar{\beta}}{2(c+1)}\int_0^1 (1-x^2)^{c+1}\,dx$

The last equality is obtained integrating by parts :

Moreover for any $d > 0$

$$\int_0^1 (1-x^2)^{d+1}\,dx = \int_0^1 (1-x^2)^d dx - \int_0^1 x^2(1-x^2)^d\,dx$$

$$= \int_0^1 (1-x^2)^d dx - \frac{1}{2(d+1)}\int_0^1 (1-x^2)^{d+1}\,dx$$

so that $\int_0^1 (1-x^2)^{d+1}\,dx = \frac{2d+2}{2d+3}\int_0^1 (1-x^2)^d\,dx$

This gives

$$A' = \frac{2c+1}{2c}\int_0^1 (1-x^2)^c\,dx \quad \text{and} \quad C' = -\frac{\bar{\beta}}{2c+2}\cdot\frac{2c+2}{2c+3}\int_0^1 (1-x^2)^c\,dx$$

And hence $S(\lambda') = 0$ is equivalent to :

$$\nu' = A\mu_1' + C$$

with :

$$A = \frac{2c+1}{2c\beta} < 0, \qquad C = -\frac{1}{2c+3} < 0 \tag{36}$$

This line cuts the line H since

$$|A| = \left(\frac{2c+1}{2c}\right)\cdot\frac{4}{(b^2+8)^{1/2}-b} = \left(\frac{2c+1}{2c}\right)\cdot\frac{1}{2}\cdot\left((b^2+8)^{1/2}+b\right) > b \tag{37}$$

The point of intersection is given by :

$$\mu'_1 = -\frac{C}{A+b} \quad (\text{and } \nu' = -b\mu'_1) \tag{38}$$

Comparison of this μ'_1-value with the μ'_1 -value of the point DH (Hopf of codimension 2), i.e. : $\mu'_1 = -\frac{1}{3b}$ gives :

$$-\frac{1}{3b} < -\frac{C}{A+b} \tag{39}$$

In fact, (39) is equivalent to :

$$b(1+3|C|) < |A|$$

$$\Leftrightarrow b\left(1 + \frac{3}{2c+3}\right) < \frac{2c+1}{2c|\beta|} = \frac{2c+1}{2c} \cdot \frac{1}{2} \cdot (b + (b^2+8)^{1/2})$$

$$\Leftrightarrow ((b^2+8)^{1/2}+b)\left(\frac{2c+1}{2c}\right) - 2b\left(\frac{2c+6}{2c+3}\right) > 0$$

which is equivalent to

$$e.\left(\frac{e^2+4}{e^2}\right) - 2b\left(\frac{e^2+24}{e^2+12}\right) > 0 \text{ where } c = \frac{e^2}{8} \text{ and } e = (b^2+8)^{1/2}+b = 4\bar{\alpha}$$

This is equivalent to

$$0 < (e^2+12)(e^2+4) - 2be(e^2+24)$$

$$= e^4 + 16e^2 + 48 - 2be^3 - 48be$$

$$= [(8b^4 + 64b^2 + 64) + (8b^3 + 32b)(b^2+8)^{1/2}] + 16[(2b^2+8) + 2b(b^2+8)^{1/2}]$$

$$+ 48 - 2b[(4b^2+8)(b^2+8)^{1/2} + (4b^3+24b)] - 48b[b+(b^2+8)^{1/2}]$$

$$= 240$$

3.2 The line SC_i (inferior saddle connections)

As for the curve SC_s, the limit equation of CS_i is given by $I(\lambda') = 0$, where $I(\lambda')$ is the following integral, obtained taking $r = \frac{1}{\alpha}$:

$$I(\lambda') = \int_{-1}^{1} [(\bar{\beta}-\bar{\alpha})\ (x^2-1)]^{-(\frac{\bar{\beta}}{\alpha}+1)} (\mu_1'+\nu'\bar{\alpha}(x^2-1) + \bar{\alpha}x^2(x^2-1))\ dx$$

$$= (\bar{\alpha}-\bar{\beta})^{-(\frac{\bar{\beta}}{\alpha}+1)} [(\int_{-1}^{1} (1-x^2)^{\frac{1}{c}-1}\ dx)\ \mu_1' - (\bar{\alpha} \int_{-1}^{1} (1-x^2)^{\frac{1}{c}}\ dx)\ \nu'$$

$$- (\bar{\alpha} \int_{-1}^{1} x^2(1-x^2)^{\frac{1}{c}}\ dx)] \tag{40}$$

where again $c = \frac{1}{8}\ (b + (b^2+8)^{1/2})^2 > 1$ and $\bar{\alpha} = \frac{1}{4}\ (b + (b^2+8)^{1/2})$

Calculations similar to those performed in the case SC_s lead to the fact that $I(\lambda') = 0$ is equivalent to :

$$\nu' = \bar{A}\mu_1' + \bar{C}$$

with : $\bar{A} = \frac{2+c}{2\alpha} > 0$ and $\bar{C} = -\frac{c}{3c+2} < 0$ (41)

This line cuts the line H at the point :

$$\mu_1' = -\frac{\bar{C}}{\bar{A}+b} \quad \text{and} \quad \nu' = -b\mu_1' \tag{42}$$

(So $\mu_1' > 0$ and $\nu' < 0$).

It also cuts the line SC_s at the point TSC of coordinates :

$$\begin{cases} \mu_1^d = \frac{C-\bar{C}}{\bar{A}-A} = \frac{2}{3} \cdot \frac{\bar{\alpha}(2\bar{\alpha}^2-1)}{(2\bar{\alpha}^2+1)(3\bar{\alpha}^2+1)(4\bar{\alpha}^2+3)} > 0 \\ \\ \nu^d = -\frac{(6c^3+13c^2+7c+4)}{3(c+1)(3c+2)(2c+3)} < 0 \end{cases} \tag{43}$$

(recall $\bar{\alpha} = \frac{1}{4}((b^2+8)^{1/2}+b)$ and $c = 2\bar{\alpha}^2$)

One can easily prove, by looking to $d\nu/dc$, that ν^d is increasing for $b > 0$. For $b \to 0$ the point TSC $= (\mu_1^d, \nu^d) \to (0, -\frac{1}{5})$. For $b \to \infty$ this point converges towards $(0, -\frac{1}{3})$.

At the point TSC, there exists a double connection, i.e. a singular cycle containing the two separatrices γ_s, γ_i. By an implicit function argument there exists a line TSC (τ), with TSC(0) = TSC of such singular cycles in $\mathbb{R}^3(\tau, \mu_1', \nu')$. In the next paragraph we prove the genericity of this line. The results obtained in this paragraph are summarized in the following figure :

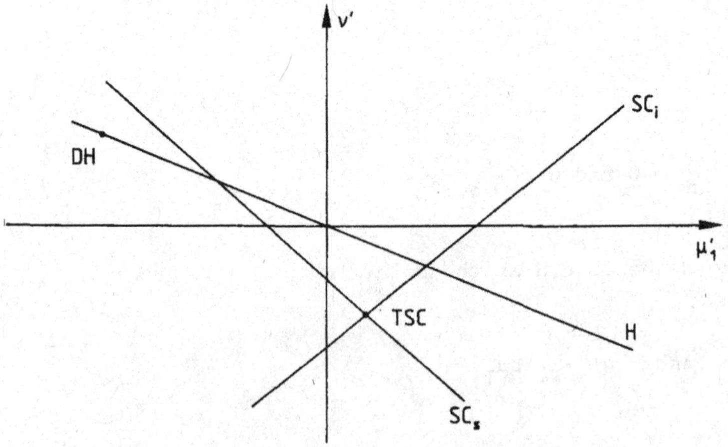

Fig. 51

4. Bifurcation point of two saddle connections (TSC)

At the end of the previous paragraph we have found the point TSC (μ_1^d, ν^d) corresponding to a terminating point of a line of two saddle connections. It remains to prove its genericity. First, notice that the situation corresponds to the degenerate case of IV.3.7. The value r_1 is

$$r_1 = \frac{\bar{\alpha}}{-\bar{\beta}} = \frac{b + (b^2+8)^{1/2}}{(b^2+8)-b^{1/2}} > 1 \tag{44}$$

Next, the equations for the surfaces SC_s, SC_i of saddle connections are given by :

$$\begin{cases} i(\lambda',\tau) = \tau I(\lambda') + o(\tau) \\ s(\lambda',\tau) = \tau\, S(\lambda') + o(\tau) \end{cases} \tag{45}$$

where $I(\lambda')$ and $S(\lambda')$ are the integrals studied in the preceeding paragraphs. The genericity of these two functions is equivalent to the transversality of the equations $\{I=0\}$ and $\{S=0\}$, established above to define the point TSC (Formula 43). So it remains to verify the conditions on the asymptotic developments of the eigenvalue ratios at the points s_1, s_2 for $\tau \to 0$.

Let $r_1(\tau) = \dfrac{\xi_1(\tau)}{\lambda_1(\tau)}$ and $r_2(\tau) = \dfrac{\xi_2(\tau)}{\lambda_2(\tau)}$ where $\lambda_1(\tau)$, $-\xi_1(\tau)$ and $\lambda_2(\tau)$, $-\xi_2(\tau)$ are the eigenvalues at s_1 and s_2 respectively, for the value τ.

Let $r_1(\tau) = r_1 + \tau\alpha_1 + o(\tau)$ and $\dfrac{1}{r_2(\tau)} = r_1 + \tau\alpha_2 + o(\tau)$.

The genericity condition is $\alpha_1 - \alpha_2 \neq 0$ (see IV.3.7).

Notice that $r_1(\tau).r_2(\tau) = 1 + \dfrac{\alpha_1-\alpha_2}{r_1}\tau + o(\tau)$

So it suffices to expand $r_1(\tau).r_2(\tau) = \dfrac{\xi_1(\tau)}{\lambda_1(\tau)} \cdot \dfrac{\xi_2(\tau)}{\lambda_2(\tau)}$

First, we compute the eigenvalues $\lambda_i(\tau)$, $-\xi_i(\tau)$, $i = 1,2$.

The singular points $s_1(\tau)$, $s_2(\tau)$ are given by : $y = 0,\ x^3 - x + \tau\mu_1' = 0$ (46)

Near (1,0), we introduce $u = x-1$.

$$u^3 + 3u^2 + 2u + \tau\mu_1' = 0$$

Around $u = 0$, this gives : $u = -\frac{1}{2}\mu_1'\tau + O(\tau^2)$, (47)

hence $x = 1 - \frac{1}{2}\mu_1'\tau + O(\tau^2)$ for the point $s_2(\tau)$ (48)

The 1-jet at this point has the following matrix :

$$\begin{pmatrix} 0 & 1 \\ 2-3\mu_1'\tau+O(\tau^2) & b+\tau(1+\nu'-\frac{b}{2}\mu_1')+O(\tau^2) \end{pmatrix} \tag{49}$$

whose eigenvalues $-\xi_2$, λ_2 are

$$\frac{1}{2}\left[b + \tau(1+\nu' - \frac{b}{2}\mu_1') \pm (b^2+8)^{1/2}\left(1 + \frac{\tau}{b^2+8}\left(b+b\nu' - \mu_1'(6+\frac{b^2}{2})\right)\right)\right] + O(\tau^2) \tag{50}$$

Near the point (-1,0), we put $u = x+1$ and obtain for the singular point

$s_1(\tau)$: $x = -1 - \frac{1}{2}\mu_1'\tau + O(\tau^2)$ (51)

and for the eigenvalues $-\xi_1$, λ_1 :

$$\frac{1}{2}\left[-b+\tau(1+\nu' - \frac{b}{2}\mu_1') \pm (b^2+8)^{1/2}\left(1 - \frac{\tau}{b^2+8}\left(b+b\nu' - \mu_1'(6+\frac{b^2}{2})\right)\right)\right] + O(\tau^2) \tag{52}$$

From (50), we obtain :

$$\frac{\xi_2}{\lambda_2} = \frac{(b^2+8)^{1/2} - b + \tau(-1-\nu' + \frac{b}{2}\mu_1' + (b^2+8)^{-1/2}(b+b\nu'-\mu_1'(6+\frac{b^2}{2}))) + O(\tau^2)}{b+(b^2+8)^{1/2} + \tau(1+\nu'-\frac{b}{2}\mu_1' + (b^2+8)^{-1/2}(b+b\nu'-\mu_1'(6+\frac{b^2}{2}))) + O(\tau^2)}$$

$$= 8\,(b+(b^2+8)^{1/2})^{-2} + O(\tau)$$

which is less than 1 for τ sufficiently small and (μ_1',ν') near (μ_1^d,ν^d).

$$\frac{\xi_1}{\lambda_1} = \frac{(b^2+8)^{1/2} + b + \tau(-1-\nu' + \frac{b}{2}\mu_1' - (b^2+8)^{-1/2}(b+b\nu'-\mu_1'(6+\frac{b^2}{2}))) + O(\tau^2)}{(b^2+8)^{1/2} - b + \tau(1+\nu'-\frac{b}{2}\mu_1' - (b^2+8)^{-1/2}(b+b\nu'-\mu_1'(6+\frac{b^2}{2}))) + O(\tau^2)} \qquad (54)$$

$$= \frac{1}{8}\,(b + (b^2+8)^{1/2})^2 + O(\tau)$$

which is greater than 1 for τ sufficiently small and (μ_1',ν') near (μ_1^d,ν^d).

Next :

$$\frac{\xi_1}{\lambda_1} \cdot \frac{\xi_2}{\lambda_2} = 1 - (b\mu_1' + 4\nu' + 4)\,\tau(b^2+8)^{-1/2} + O(\tau^2) \qquad (55)$$

As $\mu_1^d > 0$ and $\nu^d > -\frac{1}{3}$, this implies that $\alpha_1 - \alpha_2 < 0$ (56), which means that we are in the weak expanding case ($r_1 > 1$ and $\alpha_1 - \alpha_2 < 0$).

This finishes the proof of the genericity of the point TSC.

The relative position of the lines $L_r(\tau)$, $L_\ell(\tau)$, and $SC_s(\tau)$, $SC_i(\tau)$ on an arbitrary compact K of the (μ_1',ν')-space will be established in next sections. As a consequence will follow that, for τ small enough, there exists a triangular region $T(\tau)$ bounded by the lines $H(\tau)$, $L_\ell(\tau)$ and the arc $DC(\tau)$, where exactly two limit cycles coexist. (Compare with the conclusion of Paragraph 2 and Figure 50.)

5. Complete analysis of the saddle case in a large central rescaling chart.

5i. For any domain A in parameter space (μ_1', μ_2', ν') the results of paragraph 2 are valid for all parameter values in A, but they only concern the behaviour of X_λ', on some neighborhood B of (0,0) in (x',y')-space. This B may be chosen arbitrarily but has to lay inside the domain D defined by the two saddle connections (see fig. 52)

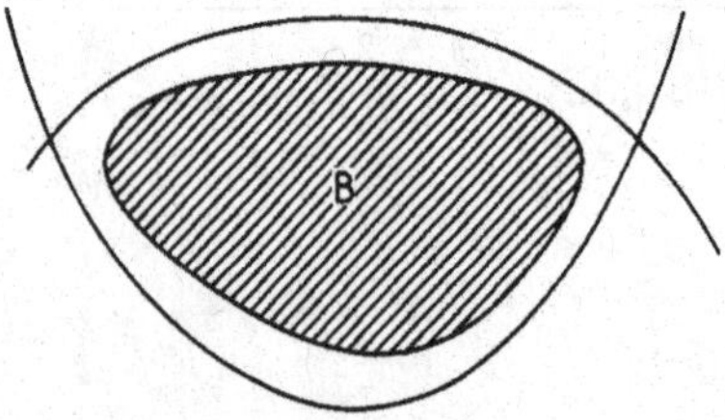

Fig. 52

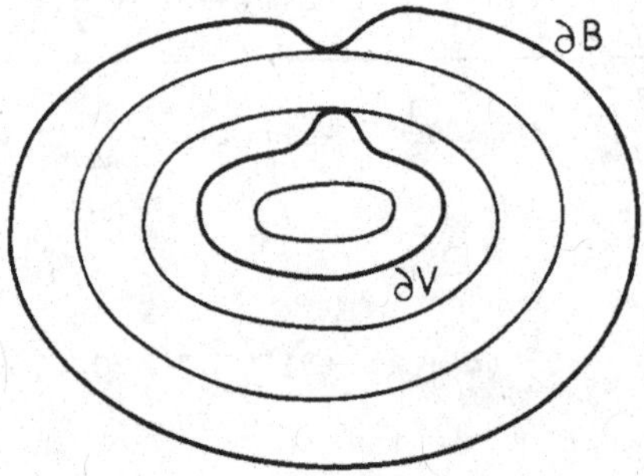

Fig. 53

The analysis in paragraph 2 hence applies to the study of closed orbits of "small and medium size" in the central chart. However in order to study the bifurcation of the saddle connections and the closed orbits near these connections - which we call the "large" limit cycles - we are now going to work in a neighbourhood of the singular cycle defined by the two connections. We will find that the complete bifurcation diagram is valid in a sufficiently small neighbourhood V of the two connections, for all parameter values in A.

To make the link with the results of paragraph 2 we first take an open set such that $\partial V \cap D$ is a level curve of ω_S; we choose a neighbourhood B of (0,0) for which ∂B is a level curve of ω_S and $\partial B \subset V$. For r sufficiently small and any $\lambda' \in A$, all closed orbits will be either in V or in B and can hence be detected.

Small and medium limit cycles

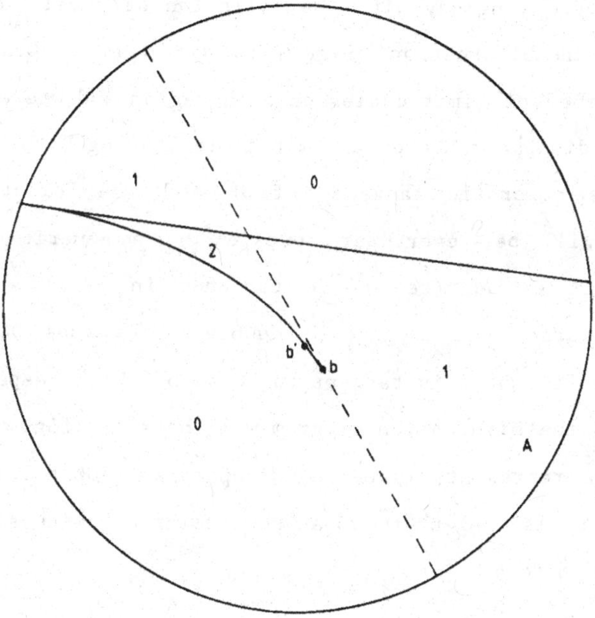

Large limit cycles

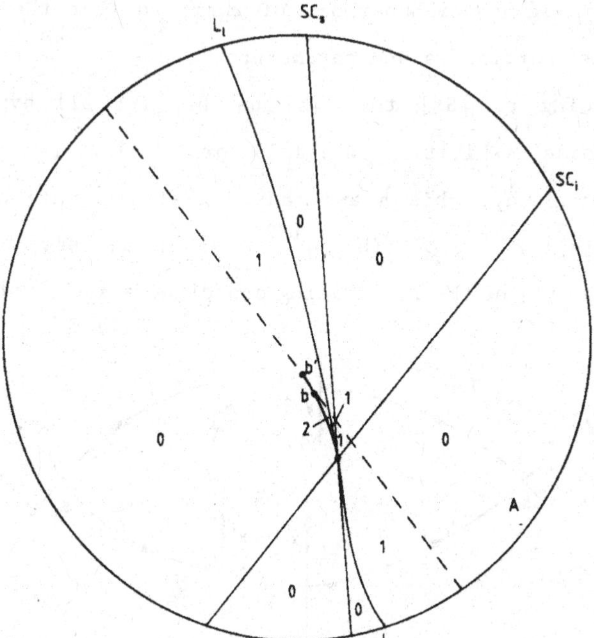

Figure 54.

To establish this link in a clear way we will modify somewhat our neighbourhoods in the way presented in fig. 53, where both ∂B and ∂V have a point of quadratic tangency with the foliation of level curves of ω_S. With respect to B the bifurcation diagram is as in fig. 54.a, where we make precise the number of limit cycles on each region and where the dashed line represents the disappearance of a limit cycle leaving B through ∂B.

Because of the quadratic tangency of ∂B with the foliation of ω_S, this disappearance will be, everywhere outside b, a generic disappearance of hyperbolic limit cycles (see cycle tangency in IV.2). In b we have a generic disappearance of a generic double cycle (see DCT in IV.3). The dashed line in fig. 54.a is tangent to DC at b. With respect to V, we will now prove that the bifurcation diagram will be as in fig. 54.b, where again the dashed line represents (generic) disappearance of a limit cycle leaving V through ∂V; it is tangent to DC at some point b', with similar properties as b with respect to B.

5.ii As the saddle points s_1, s_2 have non vanishing divergence, as $\tau \to 0$, we will assume to simplify calculations that there is a C^2- linearization of the field $X_{\tau,\lambda}$, in a neighbourhood of them, so that the charts depend in class C^2 on the coordinates and parameters.

In fact according to [St] this is possible for all hyperbolic saddles, except for those exhibiting a (1:1) or a (1:2)-resonance. For the situation under study, this means that a C^2-linearization is permitted if one excludes $((b^2+8)^{1/2} - b)\,((b^2+8)^{1/2} + b)^{-1} = \frac{1}{2}$, hence b = 1.

Let (x,y) and (x',y') be the linearizing coordinates as illustrated in fig. 55.

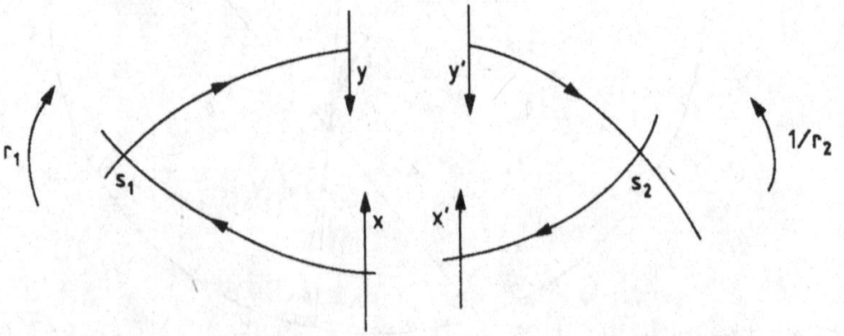

Figure 55.

Denote by r_1 and $1/r_2$ the modulus of the ratios of eigenvalues, in the sense of the arrows, as in the previous paragraph.

Let us first discuss the generic case, where we assume that $1 < r_1 < \frac{1}{r_2}$.

The transition mappings along the hyperbolic sectors can be written as $s_1 : x \to y = x^{r_1}$, $s_2 : x' \to y' = (x')^{1/r_2}$.

The transition mappings along the separatrices, which are C^2-diffeomorphisms, can be written as follows :

$$\begin{aligned} y \to y' &: y' = s + \phi(y);\ \phi(y) = v_1 y + \ldots \\ x \to x' &: x' = -i + \psi(x);\ \psi(x) = u_1 x + \ldots \end{aligned} \tag{0}$$

where $v_1 > 0$, $u_1 > 0$ and s and i measure the normal distance between the separatrices as illustrated in Fig. 56.

Here the functions s, i, ϕ, ψ also depend on class C^2 on the parameters of the system, which we have omitted to simplify the writing.

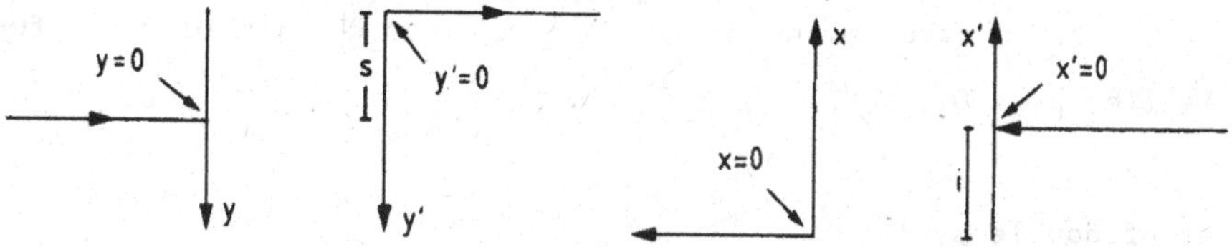

Figure 56.

To write the equation of the cycles, we equate the results of the following compositions of transition mappings :

(1) $x \to y \to y' = s + \phi(x^{r_1})$,

(2) $x \to x' \to y' = (-i + \psi(x))^{1/r_2}$, and obtain :

$$E(x) = s + \phi(x^{r_1}) - (-i + \psi(x))^{1/r_2} = 0 \tag{E}$$

The equation of L_ℓ, the loops at s_1, through $x=0$ is

$$s = (-i)^{1/r_2}, \quad i \le 0.$$

Notice that this equation does not depend on ϕ, ψ.

The equation of L_r , the loops at s_2, through $x'=0$, is

$$\psi(x) = i, \qquad s = -\phi(x^{r_1}),$$

Inverting the C^2 diffeomorphism ψ, we obtain

$$s = -\phi[(\psi^{-1}(i))^{r_1}]$$

Further development of this expression leads to

$$s = -[v_1(u_1^{-1})^{r_1}]\, i^{r_1}[1 + O(i)],$$

where $O(i)$ has a development in i^{mr_1+n}, $n, m \in \mathbb{N}$, modulo a C^2 function which is flat at $x=0$.

The line of double cycles

The double cycles are given by the following system

$$\begin{aligned} &s + \phi(x^{r_1}) = (-i + \psi(x))^{1/r_2} \\ &r_1\, \phi'(x^{r_1})\, x^{r_1-1} = \frac{1}{r_2}\, (-i + \psi(x))^{\frac{1}{r_2}-1}\, \psi'(x) \end{aligned} \qquad (E')$$

Evaluating $-i + \psi(x)$ from second equation, we obtain

$$-i + \psi(x) = [r_1 r_2 \frac{\phi'(x^{r_1})}{\psi'(x)}\, x^{r_1-1}]^{1/(1/r_2-1)}.$$

Therefore, the parametric equation of the line of double cycles becomes

$$-i = \left[r_1 r_2 \frac{\phi'(x^{r_1})}{\psi'(x)} x^{r_1 - 1}\right]^{r_2/1-r_2} - \psi(x)$$

$$s = \left[r_1 r_2 \frac{\phi'(x^{r_1})}{\psi'(x)} x^{r_1 - 1}\right]^{1/1-r_2} - \phi(x^{r_1})$$

Further calculation taking into account that $1 < r_1 < 1/r_2$ gives that, asymptotically, $-i \sim s^{r_2}$, which amounts to $s \simeq (-i)^{1/r_2}$.

Therefore, the line of double cycles and that of left loops, L_ℓ, cannot be distinguished at this level of analysis. A qualitative argument places the line of loops below that of double cycles, as illustrated in Fig. 57. The reader is invited to compute higher order terms in the asymptotic developments of the double cycle line to verify this part analytically.

Before sketching the qualitative argument, we observe that outside the line DC all closed orbits are hyperbolic, while the line of double cycles is a line of generic double cycle bifurcations. Indeed a calculation along this line of the second derivative $\frac{d^2E}{dx^2}$ for $x \neq 0$ shows that it cannot be zero there.

The boundary of the neighbourhood V of the 2 connections may (even with V arbitrarily small) be chosen tranverse to the flows, with the vector fields pointing outwards.

As the left loops are attracting, there is the creation of an attracting limit cycle; for fixed s we clearly need to find these limit cycles for increasing i. Between this attracting limit cycle and ∂V there must be at least one expanding limit cycle.

On $\{i=0,\ s>0\}$ there is no closed orbit since (E) has no solution there for x sufficiently small. Hence for each fixed s (>0) and i increasing, the two limit cycles have to coalesce.

To complete the picture we observe that along L_r there is creation of an expanding limit cycle in the direction of decreasing i, when $s<0$ remains constant. The number of limit cycles is hence exactly as it is presented in fig. 57.

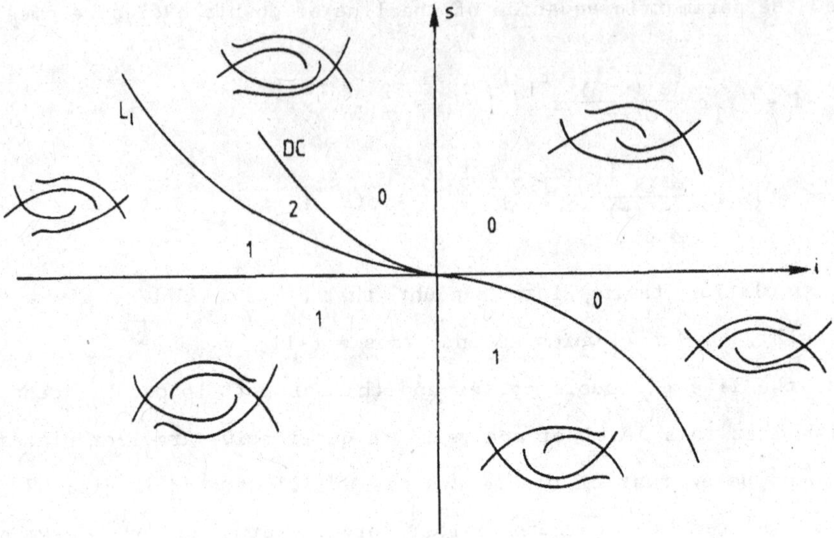

Figure 57.

5.iii Analysis of the Hamiltonian Perturbation.

A. Preliminaries

In paragraph 3 we worked with the parameters τ, μ_1', ν', where τ is small. We would now like to obtain more information about the bifurcation diagram around TSC in the plane (μ_1',ν'), as $\tau\to 0$. We have C^∞ dependence, with respect to (μ_1', ν', τ) of the functions $i = i(\mu_1', \nu', \tau)$, $s = s(\mu_1', \nu', \tau)$ which measure the normal distance between the separatrices.
The functions $I(\mu_1',\nu') = \frac{\partial i}{\partial \tau} \mid (\tau=0)$, $S = \frac{\partial s}{\partial \tau} \mid (\tau=0)$, are of class C^∞ and, in fact, linear relative to (μ_1',ν'). The map $(\mu_1',\nu') \to (I,S)$ is a global C^∞ diffeomorphism.

Denote by x_1, y_1; x_1', y_1' the linearizing coordinates introduced above, where one coordinate system is valid for a neighbourhood of each saddle. Denote also by x,y; x',y' the coordinates defined by the normal forms. The coordinate changes being of class C^2. Therefore, if i_1, s_1 are the new normal separation functions associated to the coordinates (x_1,y_1,x_1',y_1'), the coordinate changes induce C^2 diffeomorphisms on the transversals :

$x \to x_1(x, \mu_1', \nu', \tau)$, $y' \to y_1'(y', \mu_1', \nu', \tau)$, which depend on class C^2 on the parameters.

Also :

$$i_1 = x_1(i, \mu_1', \nu', \tau), \quad s_1 = y_1'(s, \mu_1', \nu', \tau).$$

Now, the new functions I_1, S_1 satisfy :

$$I_1 = \partial i_1/\partial\tau \,|_{\tau=0} = (\partial x_1/\partial x)(\partial i/\partial\tau)\,|_{\tau=0} + \partial x_1/\partial\tau\,|_{\tau=0}$$

$$= (\partial x_1/\partial x)\, I + (\partial x_1/\partial\tau)\,|_{\tau=0},$$

$$S_1 = (\partial y_1'/\partial y')\, S + (\partial y_1'/\partial\tau)\,|_{\tau=0}.$$

Due to the preservation of invariant manifolds under changes of coordinates, we have $x_1(0, \mu_1', \nu', \tau) \equiv 0$ and $y_1'(0, \mu_1', \nu', \tau) \equiv 0$, which implies that $(\partial x_1/\partial\tau)\,|_{\tau=0} = (\partial y_1'/\partial\tau)\,|_{\tau=0} = 0$.

Furthermore, the form ω_S in paragraph 2, at $\tau = 0$, is independent on μ_1', ν'. Therefore we can suppose that the coordinate changes were chosen so that $x_1 = x_1^o(x) + 0(\tau)$, $y_1' = y_1'^o(y') + 0(\tau)$ and that $\partial x_1/\partial x$, $\partial y_1'/\partial y'$ are positive constants.

In conclusion, it can be said that the functions $I_1(\mu_1', \nu')$, $S_1(\mu_1', \nu')$ which come from the C^2 linearizations are proportional to I and S : $I_1 = aI$, $S_1 = bS$, $a, b > 0$.

Therefore, I_1, S_1 depend on class C^∞ on μ_1', ν' and we can take them as independent parameters, instead of μ_1', ν'. This expresses the transversality hypothesis, verified by the family under consideration.

Omitting from now on the sub indices, we can write that i, s are functions of class C^2 on τ, I, S,

$$i = \tau I + o(\tau)$$

$$s = \tau S + o(\tau).$$

B. Study of the line L_ℓ in the p.h. case

We intend to verify the following. On an arbitrary neighbourhood K of (0,0) in the space I,S and for τ small enough, the line $L_\ell(\tau)$ is a graph $S=S_\tau(I)$, for $I \le 0$, such that $S_\tau(I) \to 0$ in the C^1 topology, as $\tau \to 0$, and this uniformly on $I \in [-I_o, -I_1]$ for $-I_o < -I_1 < 0$.

We fix an interval $[-I_o, -I_1]$ with $-I_o < -I_1 < 0$. We know that, for $\tau \ne 0$, $L_\ell(\tau)$ has an equation of the form $s = (-i)^{1/r_2}$. As $(s,i) \to 0$ with $\tau \to 0$, the domain where this expression holds will contain the neighbourhood K, provided τ is small enough. We will use the expressions :

$$1/r_2 = r + \alpha_2 \tau + o(\tau),\ \alpha_2 = \alpha_2(I,S)$$

$$s = \tau S + o(\tau),\ i = \tau I + o(\tau).$$

The equation for $L_\ell(\tau)$ writes as follows :

$$\tau[S + O(\tau)] = [\tau(-I+O(\tau))]^{r+\alpha_2\tau+o(\tau)}$$

or

$$S + O(\tau) = \tau^{r-1+\alpha_2\tau+o(\tau)}(-I+O(\tau))^{r+\alpha_2\tau+o(\tau)}$$

Here the functions $O(\tau)$ are of class C^1 with respect to τ, I, S and the functions $o(\tau)$ are of class C^2.

Since α_2 as well as the functions $O(\tau)$ and $o(\tau)$ depend on S, the equation above defines $S(I,\tau)$ as an implicit function. We must show that $S(\tau,I)$ and $(\partial S/\partial I)(\tau,I)$ tend to zero as $\tau \to 0$, uniformly in $[-I_o, -I_1]$.

To this effect, write

$$S = \tau^{r-1+\alpha_2\tau+\psi}\,[-I+\phi_1]^{r+\alpha_2\tau+\psi} + \phi_2,$$

where ϕ_1, ϕ_2 are C^1 and ψ is C^2.

By continuity, it follows that $\partial\phi_{1,2}/\partial I$, $\frac{\partial\phi_{1,2}}{\partial S} \to 0$, as $\tau \to 0$. Also, that $\partial\psi/\partial I$, $\frac{\partial\psi}{\partial S} = o(\tau)$.

This implies, using the fact that $r>1$, that the partial derivatives with respect to I and S of the right hand member of the last equation are continuous in (τ, I, S) and go to zero as $\tau \to 0$.

From this follows, by the C^1 continuity of implicit functions on parameters, that both S and $\partial S/\partial I$ tend to zero uniformly on the compact interval $[-I_0, -I_1]$, as $\tau \to 0$.

In the same way one can observe that the lines of superior and inferior saddle connections SC_s^τ and SC_i^τ are defined as graphs $S = \phi_\tau(I)$ and $I = \psi_\tau(S)$ which both tend in a C^1 way to respectively $\{S=0\}$ and $\{I=0\}$, and this for any compact interval in the I-axis, resp. S-axis; indeed their equation is respectively

$$0 = s = \tau(S + 0(\tau))$$

$$0 = i = \tau(-I+0(\tau))$$

Because of this C^1-tendancy to the I- or the S-axis and using the fact that the line of Hopf bifurcations cuts the I-axis and the S-axis exactly once (see paragraphs 1 to 3), we see that inside K - for K sufficiently large - the line of Hopf bifurcations is going to cut L_ℓ , SC_s and SC_i in exactly 1 point (see fig. 54).

C. Study of the equation for double cycles.

For the sake of simplicity we will assume that in expression (0) of 5.ii, $\phi(x) = x$ and $\psi(x) = x$ and also that $s = \tau S$, $i = \tau I$.

The equation for cycles is as follows : $\tau S + x^{r_1} = (-\tau I+x)^{1/r_2}$, where we take $r_1 = r + \tau\alpha_1$, $1/r_2 = r + \tau\alpha_2$.

It follows that the equation of cycles is

$$\tau S + x^{r+\tau\alpha_1} = (-\tau I+x)^{r+\tau\alpha_2} \tag{56}$$

The equation for double cycles is obtained by eliminating x from (56) and its derivative given by :

$$(r + \tau\alpha_1)\, x^{r-1+\tau\alpha_1} = (r+\tau\alpha_2)\, (-\tau I+x)^{r-1+\tau\alpha_2} \tag{57}$$

For $\tau\neq 0$ equations (56) and (57) reduce to :

$$S + \frac{1}{\tau}\, x^{r+\tau\alpha_1} - \tau^{r-1+\tau\alpha_2} (-I + \frac{x}{\tau})^{r+\tau\alpha_2} = 0 \tag{58}$$

$$(\frac{r}{\tau} + \alpha_1)\, x^{r-1+\tau\alpha_1} - (\frac{r}{\tau} + \alpha_2)\, (-\tau I+x)^{r-1+\tau\alpha_2} = 0 \tag{59}$$

This can be transformed into

$$S = \frac{1}{\tau} [(\frac{r+\tau\alpha_1}{r+\tau\alpha_2})^{\frac{r+\tau\alpha_2}{r-1+\tau\alpha_2}} x^{\frac{r-1+\tau\alpha_1}{r-1+\tau\alpha_2} \cdot (r+\tau\alpha_2)} - x^{r+\tau\alpha_1}] \tag{60}$$

$$I = \frac{1}{\tau} [x - (\frac{r+\tau\alpha_1}{r+\tau\alpha_2})^{\frac{1}{r-1+\tau\alpha_2}} . x^{\frac{r-1+\tau\alpha_1}{r-1+\tau\alpha_2}}] \tag{61}$$

From (61) :

$$I = \frac{1}{\tau} [x - (1 - \frac{\tau}{r(r-1)} (\alpha_2-\alpha_1) + 0(\tau^2))\, x^{1 - \frac{\tau(\alpha_2-\alpha_1)}{r-1} + 0(\tau^2)}] \tag{62}$$

The limiting equation for $\tau \to 0$ can be obtained by looking at the limit for $\tau \to 0$ of the right hand side of (62), which is :

$$\frac{\partial}{\partial\tau} ((1 - \frac{\tau}{r(r-1)} (\alpha_2 - \alpha_1) + 0(\tau^2))\, x^{1 - \frac{\tau(\alpha_2-\alpha_1)}{r-1} + 0(\tau^2)}) \Big|_{\tau=0}$$

$$= \frac{\alpha_2-\alpha_1}{r-1} x \ln x + \frac{\alpha_2-\alpha_1}{r(r-1)} x$$

From (60) :

$$S = \frac{1}{\tau} [(1 - \frac{\tau}{r} (\alpha_2-\alpha_1) + 0(\tau^2))^{\frac{r}{r-1}(1 + \frac{\tau\alpha_2}{r(r-1)} + 0(\tau^2))}$$

$$\times x^{(1 - \frac{\tau}{r-1} (\alpha_2-\alpha_1) + 0(\tau^2))(r+\tau\alpha_2)} - x^{r+\tau\alpha_1}]$$

$$= \frac{x^{\tau\alpha_1}}{\tau} [(1 - \frac{\tau}{r} (\alpha_2-\alpha_1) + 0(\tau^2))^{\frac{r}{r-1}(1 + \frac{\tau\alpha_2}{r(r-1)} + 0(\tau^2))}$$

$$\times x^{r - \frac{\tau}{r-1} (\alpha_2-\alpha_1) + 0(\tau^2)} - x^r] \tag{63}$$

To obtain the limiting equation for $\tau \to 0$. We look at $\frac{1}{\tau}$ times the expression in between brackets in the right hand side of (63), nl. :

$$\lim_{\tau\to 0} \frac{1}{\tau} [(1 - \frac{\tau}{r-1} (\alpha_2-\alpha_1) + 0(\tau^2)) x^{r - \frac{\tau}{r-1} (\alpha_2-\alpha_1) + 0(\tau^2)} - x^r]$$

$$= \frac{\partial}{\partial\tau} ((1 - \frac{\tau}{r-1} (\alpha_2-\alpha_1) + 0(\tau^2)) x^{r - \frac{\tau}{r-1} (\alpha_2-\alpha_1) + 0(\tau^2)}) \Big|_{\tau=0},$$

$$= - \frac{\alpha_2-\alpha_1}{r-1} x^r - \frac{1}{r-1} (\alpha_2-\alpha_1) x^r \ln x$$

By this the limit position of the line of double cycles is given by the parametrization :

$$I = \beta(x \ln x + \frac{1}{r} x)$$
$$S = - \beta(x^r \ln x + x^r) \tag{64}$$

where $\beta = \dfrac{\alpha_2-\alpha_1}{r-1} > 0$.

We see that $I < 0$, while $S > 0$ for $x > 0$ sufficiently small.

Also for $x \geq 0$ sufficiently small, the equations $(I(x,\tau), S(x,\tau))$ given by (56) and (57) tend in a uniform way to $(I(x,0), S(x,0))$ given by (64)).

Therefore, if we fix an interval $[-I_0, -I_1]$ with $-I_0 < -I_1 < 0$, we see that there is no double cycle for small values of x. This justifies the existence of a compact neighbourhood K′ of $[-I_0, -I_1] \times \{0\}$ in (I,S)-space, such that for $(I,S) \in K'$ there are only hyperbolic cycles.

As we have seen, using the rotational property, there is at most one cycle, which is hyperbolic for (I,S) in K′.

6. Study in a "principal rescaling cone" around the TSC-line.

Let us come back to the family obtained after principal rescaling (see I.3 and V.E)

$$\bar{y}\frac{\partial}{\partial \bar{x}} + (\bar{x}^3 + \bar{\mu}_2\bar{x} + \bar{\mu}_1 + \bar{y}(\bar{\nu} + b\bar{x})\frac{\partial}{\partial \bar{y}} + t(\bar{x}^2 + d\bar{\nu}\bar{x} + O(t))\,\bar{y}\frac{\partial}{\partial \bar{y}} \tag{1}$$

with $d = \frac{\partial b}{\partial \nu}(0)$.

We will consider the principal rescaling chart : $\{\bar{\mu}_2 = -1\}$, in which the expression of the family is :

$$\bar{y}\frac{\partial}{\partial \bar{x}} + (\bar{x}^3 - \bar{x} + \bar{\mu}_1 + \bar{y}(\bar{\nu} + b\bar{x}))\frac{\partial}{\partial \bar{y}} + t(\bar{x}^2 + d\bar{\nu}\bar{x} + O(t))\,\bar{y}\frac{\partial}{\partial \bar{y}} \tag{2}$$

In order to reduce the study of the saddle-case to the conjecture about the polynomial family X^P_{λ}, as announced in I.4, we will need to know (2) in a neighborhood of $\{t = \bar{\mu}_1 = \bar{\nu} = 0\}$. To this end, we use the following rescaling :

$$\begin{cases} \bar{x} = x' \\ \bar{y} = y' \end{cases} \qquad \begin{cases} \bar{\nu} = \tau\nu' \\ \bar{\mu}_1 = \tau\mu_1' \\ t = \tau u \end{cases} \tag{3}$$

changing (2) into :

$$y'\frac{\partial}{\partial x'} + (x'^3 - x' + bx'y')\frac{\partial}{\partial y'} + \tau(\mu_1' + \nu'y' + ux'^2y')\frac{\partial}{\partial y'} + O(\tau^2) \tag{4}$$

We recall that the composition of the principal rescaling with (3) gives exactly the central rescaling. The study which we made in the preceeding paragraphs consists in taking $u = 1$ and (μ_1', ν') in some compact region, implying the knowledge of (4) for parameter values inside a cone

$$C_1 = \{u=1, |\mu_1'| \leq K, |\nu'| \leq K, 0 < \tau \leq \tau_0\}$$

with $K > 0$ any fixed number and $\tau_o = \tau_o(K)$.

In order to make a study of (2) in a full neigborhood of $(t, \bar{\mu}_1, \bar{\nu}) = (0,0,0)$ in $\{u \geq 0\}$ we will now make a similar study inside the following two cones :

$$C_2 = \{\mu_1' = \pm 1, 0 \leq u \leq L, |\nu'| \leq L, 0 < \tau \leq \tau_o'\}$$

for some sufficiently small $L > 0$ and $\tau_o' > 0$,

$$C_3 = \{\nu' = \pm \nu_0', |\mu_1'| \leq 1, 0 \leq u \leq M, 0 < \tau \leq \tau_o''\}$$

for any sufficiently small ν_0' , some sufficiently small $M > 0$ and $\tau_o'' > 0$, both possibly depending on ν_0' .

Clearly, taking C_2 first, making ν_0' sufficiently small and then choosing K sufficiently large, we obtain that $C_1 \cup C_2 \cup C_3$ is a full neighborhood of $(0,0,0)$ in (u,μ_1', ν')-space for $u \geq 0$.

i. Abelian integrals :

As we saw in VI.B.2., inside $\bar{C}_1$, for τ_o sufficiently small, the limit cycles of (4) in the complement of a small neighborhood of the double heteroclinic cycle of X_S were determined by the zeroes of the following integral :

$$J(h, \mu_1', \nu_1', u) = \mu_1' J_0 + \nu' J_1 + u J_2 \tag{5}$$

with $u = 1$.

Now we are going to study this integral inside C_2 and C_3, following what was done in VI.B.2.

Consider

$$\frac{J}{J_0} = \mu_1' - \nu' P_1 - u P_2 \tag{6}$$

where $P_i = -J_i/J_0$, $i = 1,2$.

a : Study of $\frac{J}{J_o}$ in the cone C_2.

$$\frac{J}{J_o} = \pm 1 - \nu' P_1 - u P_2 \tag{7}$$

If L is sufficiently small then this function has constant sign and hence J has no zeroes.

b : Study of $\frac{J}{J_o}$ in the cone C_3

$$\frac{J}{J_o} = \mu_1' \mp \nu_o' \left(P_1(h) \pm \frac{u}{\nu_o'} P_2(h)\right) \tag{8}$$

If $\nu_o' > 0$ is fixed, when $u \to 0$, the function $P_1(h) \pm \frac{u}{\nu_o'} P_2(h)$ tends to the function $P_1(h)$ in the C^1 sense, and this limit is uniform on $[h_o, h_1]$, for $h_o < h_1$.

As P_1 has a nonzero derivative on $[h_o, h_1]$, the function $P_1 \pm \frac{u}{\nu_o'} P_2$ will have the same property on $[h_o, h_1]$ for M suficiently small.

ii. The double heteroclinic cycle of X_S

To finish, we need to study the bifurcation of saddle connections in the cones C_2, C_3.

In the notations of §3, the heteroclinic connections SC_s and SC_i are given respectively by :

$$S(\mu_1', \nu', u) = A'\mu_1' + B'\nu' + C'u = 0 \tag{9}$$

$$I(\mu_1', \nu', u) = \overline{A}'\mu_1' + \overline{B}'\nu' + \overline{C}'u = 0 \tag{10}$$

The homoclinic connections L_ℓ and L_r are given by the formulas of §5.iii.B, where S,I are now the functions (9) and (10).

Hence, all these bifurcations (or at least their limit positions) are globally defined by homogeneous linear equations in the parameters (μ_1', ν', u).

These equations are obtained by homogenizing the associated expressions used in C_1. They keep the same properties.

The related bifurcation surfaces do not interesect the cone C_2 if L is sufficiently small; they will lay in $C_1 \cup C_3$, if ν'_o is sufficiently small.
To conclude, we find that the bifurcation diagram as given in Figure 54, whose validity was previously proved inside C_1, for K sufficiently large, reveals to be correct also inside sufficiently small neighborhoods of TSC in principal rescaling.
Precisely, the diagram Σ of Fig. 54 represents exactly the intersection of the bifurcation diagram of X_λ with small spheres $\mu_1^2 + \mu_2^2 + \nu^2 = \epsilon^2$ inside some cone $\{(t\bar{\nu}, t^3\bar{\mu}_1, -t^2), t > 0, |\bar{\nu}| \leq \delta_1, |\bar{\mu}_1| \leq \delta_2\}$.

VI.C. The elliptic case

Here $\epsilon = -1$. The case $\mu'_2 = 1$ is of no interest for the bifurcation analysis because X^S is structurally stable. (For λ' in some arbitrarily fixed compact domain).

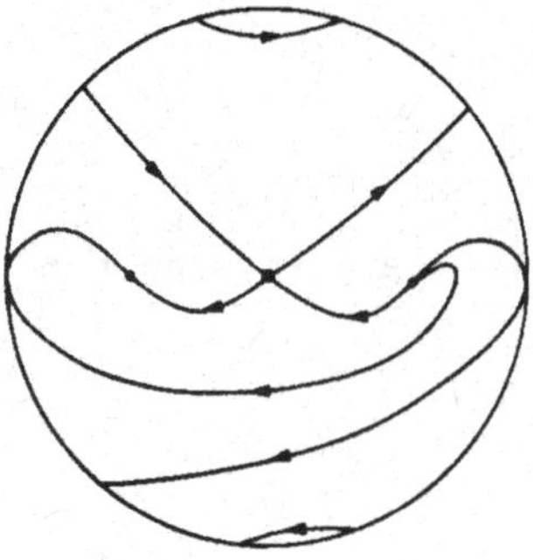

Elliptic case, $\mu'_2 = 1$

Figure 58

Suppose now that $\mu'_2 = -1$

$$\begin{cases} X_{(\lambda';\tau)} = X^S + \tau((\mu'_1 + \nu' y + x^2 y) + O(\tau)y) \dfrac{\partial}{\partial y} \\ X^S = y \dfrac{\partial}{\partial x} + (-x^3 - x + bxy) \dfrac{\partial}{\partial y} \end{cases} \qquad (65)$$

Recall that $b > 2\sqrt{2}$.

There exists just one singular point at (0,0), with eigenvalues $\pm i$ for X^S (a center). For the vector field $X_{(\lambda',\tau)}$, the position of the singular point is given by :

$$-x^3 - x + \tau\mu_1' = 0 \tag{66}$$

It follows that the x-coordinate of the critical point has the following expansion :

$$x = \tau\mu_1' + o(\tau) \tag{67}$$

The line H of Hopf bifurcations is given by (67) and :

$$bx + \tau\nu' + \tau x^2 = 0 \tag{68}$$

This gives, by elimination of x between (67), (68) :

$$\tau(b\mu_1' + \nu' + O(\tau)) = 0 , \tag{69}$$

and the following limit equation for H, when $\tau \to 0$:

$$b\mu_1' + \nu' = 0 \tag{70}$$

The candidate for the codimension 2 Hopf bifurcation (line DH in $\mathbb{R}^3$) has been calculated in V.B (formula (27))

$$\begin{cases} \nu = -bu^2 + O(u^4) \\ \mu_1 = 3bu^4 + O(u^6) \\ \mu_2 = -3bu^2 + O(u^4) \end{cases} \tag{71}$$

Take $u = \dfrac{t}{(3b)^{1/2}}$ in (71) and get, as in the saddle case, that the limit position of the line DH, when $\tau \to 0$, is the point DH $= (-\frac{1}{3}, \frac{1}{3}b)$ in the plane (μ_1', ν').

To prove the genericity of DH, for small τ, we used as in the saddle case, an integrating factor. In fact, the calculations are very similar. One has just to change b^2+8 by b^2-8, adapting some signs. So we omit the details.

Introduce (like in the saddle case but changing b^2+8 into b^2-8) :

$$\bar{\alpha} = \frac{1}{4}\left(b + (b^2-8)^{1/2}\right)$$

$$\bar{\beta} = \frac{1}{4}\left(b - (b^2-8)^{1/2}\right)$$

$$V = y - \bar{\alpha}v = y - \bar{\alpha}(x^2+1) \tag{72}$$

$$Y = \bar{\beta}v - y = \bar{\beta}(x^2+1) - y$$

We can choose :

$$K = V^{\bar{\bar{\alpha}}-1}\ Y^{\bar{\bar{\beta}}-1} \quad \text{as integrating factor,}$$

$$H = -\frac{1}{\bar{\bar{\alpha}}+\bar{\bar{\beta}}}\ V^{\bar{\bar{\alpha}}}\ Y^{\bar{\bar{\beta}}} \quad \text{as Hamiltonian,} \tag{73}$$

with $\bar{\bar{\alpha}} = r\bar{\alpha}$, $\bar{\bar{\beta}} = -r\bar{\beta}$ $\quad r \in \mathbb{R} - \{0\}$.

This allows us to make precise the phase portrait of X^S :

$\{y = \bar{\beta}(x^2+1)\}$

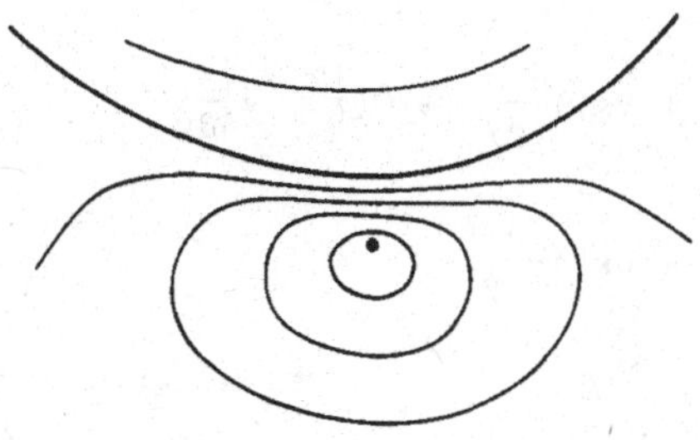

Fig. 59

The Hamiltonian used above was introduced by Zoladek in [Z1]. As in the saddle case, he performed a change in coordinates and parameters (bringing the parabola $\{y = \bar{\beta}(x^2+1)\}$ on the Ox-axis), to obtain a new Abelian integral $I = \epsilon_o I_o + \epsilon_1 I_1 + \epsilon_2 I_2$, equivalent to the one linked to the initial one (K,H). He proved the <u>local convexity</u> of the curve $Q(c) = (Q_1(c), Q_2(c))$ where $Q_1 = \frac{I_1}{I_o}$, $Q_2 = \frac{I_2}{I_o}$, at the value $c = 1$ corresponding to the center (Lemma 4.1 of [Z1]). This implies the genericity of the bifurcation point DH (Same proof as in the saddle case).

<u>Remark</u> In fact Zoladek has proved that the curve Q is globally convex. But this result seems of no use here, because we must restrict ourselves to a compact domain in the (μ_1',ν')-plane in order to apply the Perturbation Lemma. The fact that we should take any compact domain and not some compact neighborhood of DH does not modify qualitatively our result. Perhaps a global knowledge of Q would be useful to study the boundary bifurcations. But we have to keep in mind that the domain of study, $\bar{A}$, in the phase space, when we use the central rescaling, shrinks to zero with τ and does not permit, at first glance, to obtain results for the boundary ∂A of a fixed domain A in the phase space. We return to these problems in Chapter VII.

VI.D. <u>The focus case</u>

1. Study along the μ_2-axis in a large central rescaling chart.

Here we have $\epsilon = -1$ and the two cases $\mu_2' = \pm 1$ are of interest. The equation of the family is given by :

$$X_{(\lambda',\tau)} = X^S_{\pm} + \tau(\mu_1' + \nu' y + yx^2)\frac{\partial}{\partial y} + O(\tau^2)\, y\frac{\partial}{\partial y}$$

$$X^S_{\pm} = y\frac{\partial}{\partial x} + (-x^3 \pm x + bxy)\frac{\partial}{\partial y} \tag{74}$$

and $0 < b < 2\sqrt{2}$.

In the case $\{\mu_2' = 1\}$, the symmetric field X^S_+ has 3 singularities and in the case $\{\mu_2' = -1\}$, X^S_- has just one singularity (see the pictures of the phase portraits in Figure 48). Clearly, a possibility of Hopf bifurcation occurs just when $\mu_2' = -1$. Exactly as in the elliptic case above, we find that the Hopf bifurcation set tends to the line :

$$(H) : b\mu_1' + \nu' = 0 \tag{75}$$

Also, the candidate for codimension 2 Hopf bifurcation is a line tending to the point : $DH = (-\frac{1}{3}, \frac{1}{3b})$ in the (μ_1',ν')-plane. To prove that DH is a non degenerate point of bifurcation (of P.H. type) one may use again a first integral. This first integral may also be used to detect and prove the non degeneracy of the point of degenerate loop bifurcation DL, for $\mu_2' = 1$. It was introduced in [Z2] by Zoladek to study the Abelian integrals related to this focus family. To avoid a tedious transformation of normal forms, we recall calculations from [Z2], using its own notations. The equation of the orbits is given by :

$$\begin{cases} \dot{x} = y \\ \dot{y} = \sigma x + \alpha x^3 + xy + \beta_0 + \beta_1 y + \beta_2 x^2 y \end{cases} \tag{76}$$

where $\sigma = \pm 1$ and α verifies $\alpha < -\frac{1}{8}$ (α is related to b : $\alpha = -b^{-2}$). Also the parameters $\beta_0, \beta_1, \beta_2$ are related to μ_1', ν' and τ by the formulas (29) of VI.B.2.

The formulas (17) and (21) used in the saddle-case for the integrating factor K and Hamiltonian H are always valid.

But now, $\bar{\alpha}$, $\bar{\beta}$ may be replaced by eigenvalues at a focus point and are complex conjugate numbers λ, $\bar{\lambda}$:

$$\lambda = \frac{1}{2}(1 - i\beta) \text{ with } \beta = (|8\alpha+1|)^{1/2} \tag{77}$$

To obtain real functions H, K it suffices to choose r in the formulas (21) to be a purely imaginary number. Doing so, we may obtain as in Zoladek [Z2]:

$$H = \left[\frac{\sigma+\alpha x^2+\lambda y}{\sigma+\alpha x^2+\bar{\lambda} y}\right]^{i/2\beta} \left[(\sigma + \alpha x^2 + y/2)^2 + \left(\frac{\beta y}{2}\right)^2\right]^{-1/2} \tag{78}$$

This function is real. This becomes obvious if we first introduce

$$u = 1 + \sigma(\alpha x^2 + \frac{y}{2}) \text{ and } v = \sigma\beta\, y/2 \tag{79}$$

as regular coordinates in the half plane $\{x > 0\}$ and their polar coordinates $R = (u^2+v^2)^{1/2}$ and $tg(\theta) = \frac{v}{u}$ where θ is uniquely defined in $\{x \geq 0\}$ by choosing $\theta = 0$ for $u = 1$, $v = 0$ (the origin).

In the coordinates (R,θ) we have :

$$H = R^{-1} e^{\theta/\beta} \quad \text{and } K = -R^{-2} H \tag{80}$$

These functions extend naturally to analytic functions of (x,y) in all the (x,y)-plane for $\{\sigma = -1\}$ and outside the 2 node points $\{x = \pm\sqrt{-1/\alpha}\ , y = 0\}$ for $\{\sigma = 1\}$. The level $\{H = 1\}$ corresponds to the center $(0,0)$ for $\{\sigma = -1\}$ and to the loop Γ through the saddle $(0,0)$ for $\{\sigma = 1\}$. In fact we are just interested in the region $\{0 \leq H \leq 1\}$ where the levels are cycles $\Gamma_h = \{H = h\}$. This region is the whole plane for $\{\sigma = -1\}$ and the exterior of the loop $\Gamma = \Gamma_1$ for $\{\sigma = 1\}$.

Now the related Abelian integral is :

$$J(h,\beta_0,\beta_1,\beta_2) = \int_{\Gamma_h} \omega_D \quad \text{where } \omega_D = K(\beta_0 + \beta_1 y + \beta_2 x^2 y)dx \tag{81}$$

with $J_0 = \int_{\Gamma_h} Kdx$, $J_1 = \int_{\Gamma_h} Kydx$ and $J_2 = \int_{\Gamma_h} Kx^2ydx$, J is the linear combination :

$$J = \beta_0 J_0 + \beta_1 J_1 + \beta_2 J_2. \tag{82}$$

In [Z2], Zoladek has computed expressions for the J_i. For each h, $0 < h < 1$, let Γ_h^+ be the part of Γ_h in $\{x \geq 0\}$. This arc is parametrized by $\theta \in [\theta_1(h), \theta_2(h)]$ where :

$$-\pi + \theta_\beta < \theta_1(h) < 0 < \theta_2(h) < \theta_\beta \text{ for } \{\sigma = -1\} \text{ and } 0 < \theta_1(h) < \theta_\beta < \pi + \theta_\beta$$

$$< \theta_2(h) < \theta_2(1) \text{ for } \{\sigma = 1\} \text{ and } \theta_\beta = \text{Arctg}\beta.$$

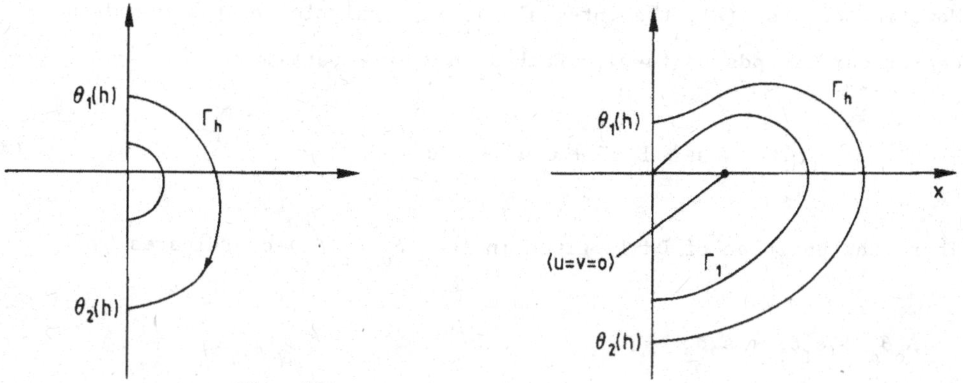

Fig. 60

Denote $z^2 = -\alpha h x^2 = \sigma(h + e^{\theta/\beta} \dfrac{\sin(\theta - \theta_\beta)}{\sin\theta_\beta})$ (83)

Then we have :

$$J_o(h) = \frac{-2\,|\alpha|^{-1/2}\sigma h^{5/2}}{\sin^2\theta_\beta}\, M_o \quad \text{with } M_o = \int_{\theta_1(h)}^{\theta_2(h)} e^{-\theta/\beta} \sin\theta \frac{d\theta}{z}$$

$$J_1(h) = \frac{4|\alpha|^{-1/2}h^{3/2}}{\beta \sin^2\theta_\beta}\, M_1 \quad \text{with } M_1 = \int_{\theta_1(h)}^{\theta_2(h)} \sin^2\theta \frac{d\theta}{z} \tag{84}$$

$$J_2(h) = \frac{4|\alpha|^{-1/2}h^{1/2}}{\beta \sin^2\theta_\beta}\, M_2 \quad \text{with } M_2 = \int_{\theta_1(h)}^{\theta_2(h)} \sin^2\theta\, z\, d\theta$$

These formulas are used to verify that

a) In the case $(\sigma = -1)$: the point DH is generic

b) In the case $(\sigma = 1)$: there exists a generic codimension 2 loop bifurcation DL. (In the P.H. sense).

We have not made in full detail the necessary calculations. Here we limit ourselves to indicate the numerical facts to prove.

To study the $J_i(h)$ for $h<1$ near 1, we introduce the variable $u = 1-h$.

Now, in the case (a), the integrals J_i are analytic for $u \geq 0$ and the center corresponds to $\{u=0\}$. Each J_i has an expansion :

$$J_i(u) = A_i u + B_i u^2 + C_i u^3 + o(u^3) \tag{85}$$

Then, the position of DH is given in the $(\beta_0, \beta_1, \beta_2)$-coordinates by :

$$\begin{aligned} A_0\beta_0 + A_1\beta_1 + A_2\beta_2 &= 0 \\ B_0\beta_0 + B_1\beta_1 + B_2\beta_2 &= 0 \end{aligned} \tag{86}$$

The genericity of this bifurcation amounts to :

$$\begin{vmatrix} A_0 & A_1 \\ B_0 & B_1 \end{vmatrix} \neq 0 \quad \text{and} \quad \begin{vmatrix} A_0 & A_1 & A_2 \\ B_0 & B_1 & B_2 \\ C_0 & C_1 & C_2 \end{vmatrix} \neq 0 \tag{87}$$

(See IV.B.1).

In the case (b), the integrals J_i are no longer analytic for $u = 0$ (corresponding to the loop), but they admit an asymptotic expansion in the functions u^j, $u^j \operatorname{Log} u$. We need 3 terms of this expansion :

$$J_i(u) = A_i + B_i u \operatorname{Log} u + C_i u + o(u) \tag{88}$$

Again, the position of DL is given by the formulas (86) and as it was proved in [DRS] the formulas (87) are still generic conditions for this bifurcation (this was recalled above in IV.B.4).

The following remark contributes to reduce a little the calculations. Notice that :

$$M_1 = -2\sigma \frac{dM_2}{du} \text{ and } M_0 = -2\sigma \frac{dM'}{du} \tag{89}$$

where $M'(u) = \int_{\theta_1}^{\theta_2} e^{-\theta/\beta} z \sin\theta \, d\theta$

Therefore we just have to find the order 4 expansion of the two integrals M_2 and M'. (The expansions of the J_i' are related to the expansions of the M_i's by formulas (84)). Notice that M_2, M' have the same form :

$\int_{\theta_1}^{\theta_2} F(\theta) z \, d\theta$, where $F(\theta)$ is an analytic function.

2. Study along the ν-axis.

i. Like in §1 we want to make a complete study in a large central rescaling chart K_1 in (x',y')-space, and this for parameter values near the ν-axis. We have to limit ourselves to parameter values in a small cone around the ν-axis; more precisely we take

$$|\mu_1| \le C|\nu|^7, \; |\mu_2| \le C|\nu|^{13/2} \tag{90}$$

C is any a priori given positive number, and the results will be valid for r sufficiently small. Moreover it seems not possible to perform the study at once, but we have to proceed in two steps. First we show that for any neighbourhood $K_0 \subset K_1$ of (0,0) in (x',y')-space the method of Hamiltonian bifurcation can be used in $K_1 \backslash K_0$. For $\nu > 0$ there will be no limit cycles in $K_1 \backslash K_0$, and for $\nu < 0$ there will be one hyperbolic expanding limit cycle. In this step we may permit

$$|\mu_1| \le C|\nu|^{5/2}, \quad |\mu_2| \le C|\nu|^2 \tag{91}$$

for any a priori given $C > 0$.

A second step will consist in studying the vector field inside a K_0 for K_0 sufficiently small. There will be no limit cycles in K_0.

ii. Study in $K_1\backslash K_0$.

Let us recall the expression of the focus-family after central rescaling :

$$y'\frac{\partial}{\partial x'} + (-x'^3 + \mu_2' x' + bx'y') + \tau(\mu_1' + \nu' y' + y'x'^2) + y'\,0(\tau^2))\frac{\partial}{\partial y'}\ , \qquad (92)$$

expressed as a family of 1-forms :

$$y'\,dy' - [(-x'^3 + \mu_2' x' + bx'y') + \tau(\mu_1' + \nu' y' + y'x'^2) + y'0(\tau^2)]\ dx' \qquad (93)$$

we put $\nu' = \pm 1$

$\mu_1' = \tau\tilde{\mu}_1$

$\mu_2' = \tau^2\tilde{\mu}_2$

$$y'dy' - (-x'^3 + bx'y')\ dx' - \tau(\pm 1 + x'^2)\ y'dx' - \tau^2(\tilde{\mu}_2 x' + \tilde{\mu}_1)dx' + 0(\tau^2)\ y'dx' \qquad (94)$$

$$= y'dy' - (-x'^3 + bx'y')dx' - \tau(\pm 1 + x'^2)\ y'dx' + o(\tau) \qquad (95)$$

We start with the case $\nu' = -1$ $(\nu < 0)$.

Using $(x',y') = (b^{-1/2}x,y)$, and $\alpha = -b^{-2}$, (95) is equivalent to :

$$ydy - (\alpha x^3 + xy)dx - \tau(-\ b^{-1/2} + b^{-3/2}\ x^2)\ ydx + o(\tau) \qquad (96)$$

$(0 < b < 2\sqrt{2} \Leftrightarrow \alpha < -\frac{1}{8})$.

In [Z2] Zoladek studies the more general expression

$$ydy - (\sigma x + \alpha x^3 + xy)dx + (\beta_0 + \beta_1 y + \beta_2 yx^2)dx + o(\|(\beta_0,\beta_1,\beta_2)\|) \qquad (97)$$

which amounts to (96) for $\sigma = 0$, $\beta_0 = 0$, $\beta_1 = \tau . b^{-1/2}$, $\beta_2 = -\tau b^{-3/2}$

In fact Zoladek considers the generic cases $\sigma = \pm 1$.
Nevertheless the formal calculations in [Z2] (writing of H, change of H into F, transformation of the integrals J into the integrals I and the Picard-Fuchs equations of the integrals I) remain valid in the case $\sigma = 0$.
Let us indicate the relevant formulas.
The Hamiltonian function is :

$$H = \left[\frac{\alpha x^2+\lambda y}{\alpha x^2+\bar{\lambda} y}\right]^{i/2\beta}\left[(\alpha x^2 + \frac{y}{2})^2 + (\beta \frac{y}{2})^2\right]^{-1/2} \tag{98}$$

where $\beta = (|8\alpha+1|)^{1/2} = \frac{(8-b^2)^{1/2}}{b}$, $\lambda = \frac{1}{2}(1-i\beta)$.

Introducing $u = \alpha x^2 + \frac{y}{2}$, $v = \frac{\beta y}{2}$, $R^2 = u^2+v^2$, $tg\theta = v/u$, (in $\mathbb{C}$-notation : $u+iv = Re^{i\theta}$) (98) is equal to :

$$H = R^{-1} e^{\theta/\beta} \tag{99}$$

The related integrating factor is

$$K = -R^{-3} e^{\theta/\beta} \tag{100}$$

The second Hamiltonian F (Formula (17) of [Z2]) is :

$$F = -a(a+1)\, y^a \left(x^2 + \frac{y}{\alpha+1}\right)$$

with $a = -\frac{1-i\beta}{1+i\beta}$.

If I_1, I_2 are the integrals associated to F in [Z2], and J_1, J_2 are the integrals

$$J_1(h) = \int_{H=h} K\, y\, dx, \quad J_2(h) = \int_{H=h} K\, y\, x^2\, dx,$$

we have (formulas (24) of [Z2]) :

$$J_1 = \sqrt{2}\, a^2 c^{\xi}\, I_1$$

$$J_2 = \sqrt{2}/3\, \frac{a^2}{\alpha}\, c^{\xi}\, I_2$$

with $\xi = \frac{i}{(1-a)\beta} - 1$ and the variable $c = h^{\frac{(1-a)\beta}{i}}$

For I_1, I_2, we have the following Picard-Fuchs system (see [Z1]) :

$$\begin{cases} 2(a+1)\, c\, I_1' = (2a+1) I_1 \\ \\ 2(a+1)\, c\, I_2' = (2a+3) I_2 \end{cases}$$

We have $Q = \frac{J_1}{J_2} = 3\alpha\, \frac{I_1}{I_2}$

From the Picard-Fuchs equation for (I_1, I_2) follows that :

$$\frac{Q'}{Q} = - \frac{1}{1+a} \cdot \frac{1}{c}$$

and hence $Q(c) = Ac^{-\frac{1}{1+a}}$ for some $A \neq 0$.

If we return to the real variable h :

$$Q(h) = Ah^{-\frac{(1-a)\beta}{i(1+a)}} = Ah$$

A is a real positive constant since $J_1(h) > 0$ and $J_2(h) > 0$.

We choose the annulus $C = K_1 \setminus K_0$ of the form

$$\{h_1 \le H \le h_2\}$$

with $0 < h_1 < \frac{1}{bA} < h_2$ (101)

H goes from $+\infty$ to 0 along any line θ=ct, when going from the origin to infinity in the (x,y)-plane. The annulus C has $\{H = h_1\}$ as outer boundary and $\{H = h_2\}$ as inner one.
With this choice of $C = K_1 \setminus K_0$ the integral

$$J(h) = \frac{1}{b^{1/2}} (J_1 - \frac{J_2}{b}) = \frac{J_2}{b^{1/2}} (Q - \frac{1}{b}) \tag{102}$$

changes its sign transversally between h_1 and h_2.

Hence for τ sufficiently small the form (95) with $\nu' = -1$ has just one limit cycle in C. This limit cycle is hyperbolic and it is expanding.

Now the case $\nu' = 1$ $(\nu > 0)$.

Similar calculations as in the case $\nu' = -1$, will lead to

$$J(h) = -\frac{J_2}{b^{1/2}} (Q + \frac{1}{b}) \tag{103}$$

and this integral remains strictly negative on any annulus $C = K_1 \setminus K_0$.
Hence for τ sufficiently small the form (95) with $\nu' = 1$ has no limit cycles in C. Orbits of the dual vector field of (95) just pass through C, coming from $\mathring{K}_0$ and going to $\mathbb{R}^2 \setminus K_1$.

iii. Study in K_0 (for K_0 sufficiently small).

As announced in (i) we will restrict (ν, μ_1, μ_2) to the cone defined by (90). This means that in (93) we put :

$$\begin{cases} \nu' = \pm 1 \\ \mu_1' = \tau^{10} \tilde{\mu}_1 \\ \mu_2' = \tau^{11} \tilde{\mu}_2 \end{cases}$$

in order to obtain $\omega'_{\tau,\tilde{\mu}_1,\tilde{\mu}_2} =$

$$y'dy' - (-x'^3 + bx'y')dx' - \tau(\pm 1 + x'^2)y'dx' - \tau^{11}(\tilde{\mu}_2 x' + \tilde{\mu}_1)dx' + O(\tau^2)\, y'dx' \tag{104}$$

We always restrict $(\tilde{\mu}_1, \tilde{\mu}_2)$ to $\max(|\tilde{\mu}_1|, |\tilde{\mu}_2|) \le C$ (with $C > 0$ a priori given) and take K_o so small that :

$$\operatorname{Sup}\{|x'|\,;\, x' \in K_0\} \le \frac{1}{2} \tag{105}$$

We use the fact that, for parameter values inside

$$\begin{aligned}\{(\nu,\mu_1,\mu_2) &= (\pm t,\, t^3\bar{\mu}_1,\, t^2\bar{\mu}_2);\, |\bar{\mu}_1| \le D,\, |\bar{\mu}_2| \le D\} \\ &= \{(\nu,\mu_1,\mu_2)\,;\, |\mu_1| \le D|\nu|^3,\, |\mu_2| \le D|\nu|^2\}\end{aligned} \tag{106}$$

with $D > 0$ sufficiently small and with $t>0$ sufficiently small, we know the focus-family in a fixed principal rescaling chart (see chapter V.E).

If this chart is sufficiently small, let us say

$$|\bar{x}| \le X \text{ and } |\bar{y}| \le Y\,, \tag{107}$$

and if $(\bar{\mu}_1,\bar{\mu}_2,t)$ are sufficiently small, let us say

$$|\bar{\mu}_1| \le M_1,\, |\bar{\mu}_2| \le M_2\,;\, |t| \le T \tag{108}$$

then $\bar{X}_{\bar{\lambda}}$, for $\bar{\lambda}$ inside the cone defined by (108), have no closed orbits inside the rectangle R defined by (107), neither can closed orbits pass through R.

In case $\nu' = +1$ (resp. $\nu'=-1$), points of ∂R have their ω-limit (resp. α-limit) outside R and their α-limit (reps. ω-limit) inside $\mathring{R}$.

The relation between principal and central rescaling in cones around the ν-axis is given by ($\nu' = \bar{\nu} = \pm 1$) :

$$t = \tau^2, \quad \begin{cases} x' = \tau \bar{x} \\ y' = \tau^2 \bar{y} \end{cases} \qquad \begin{cases} \mu_1' = \tau^2 \bar{\mu}_1 \\ \mu_2' = \tau^2 \bar{\mu}_2 \end{cases}$$

For τ sufficiently small, $(\bar{\mu}_1, \bar{\mu}_2) = (\tau^{-2}\mu_1', \tau^{-2}\mu_2') = (\tau^8 \tilde{\mu}_1, \tau^9 \tilde{\mu}_2)$ are inside the cone defined by (106).

Hence, for τ sufficiently small, the results obtained in the principal rescaling are valid for the family (104), however only in a domain R_τ with

$$R_\tau = \{(x',y') \in K_0 \; ; \; |x'| \leq \tau X, \; |y'| \leq \tau^2 Y\} \tag{109}$$

By this, for τ sufficiently small, $\omega'_{\tau,\tilde{\mu}_1,\tilde{\mu}_2}$ has no closed orbit in K_0, which pass through R_τ; and as we know from (ii) also no closed orbits can leave K_0 through points of ∂K_0.

Let us now suppose that some $\omega' = \omega'_{\tau,\tilde{\mu}_1,\tilde{\mu}_2}$ has a closed orbit γ inside $K_0 \backslash R_\tau$. We will show that this is impossible for τ sufficiently small.

We treat the case $\nu' = -1$ $(\nu < 0)$, the other one being similar.

Let H be the Hamiltonian, given in (98), (99) expressed in (x',y')-coordinates, with K as related integrating factor (see (100)) :

$$dH = K[y'dy' - (-x'^3 + bx'y')dx'] \tag{110}$$

If we write

$$\omega_D = \tau |K| [(1-x'^2 + 0(\tau)y'dx' - \tau^{10}(\tilde{\mu}_2 x' + \tilde{\mu}_1)dx'] \tag{111}$$

then, as $0 = \int_\gamma K\omega' = \int_\gamma dH - \int_\gamma \omega_D$,

necessarily $\int_\gamma \omega_D = 0$; let us take the clockwise orientation for γ. (112)

For τ sufficiently small,

$$1-x'^2 + 0(\tau) \geq \frac{1}{2} \tag{113}$$

and from (110) we find :

$$\exists C_2 > 0 : |K(x',y')| \geq C_2 \quad \text{for } (x',y') \in K_0 \tag{114}$$

$$\exists C_1 > 0 : |K(x',y')| \leq C_1 \tau^{-6} \quad \text{for } (x',y') \in K_0 \backslash R_\tau \tag{115}$$

We can decompose γ into $\gamma_+ = \gamma \cap \{y' \geq 0\}$

$\gamma_- = \gamma \cap \{y' \leq 0\}$;

γ_+ and γ_- are graphs of functions $y'_+(x')$, $y'_-(x')$; let $x'_1 < x'_2$ be the x'-coordinates of the points $\gamma \cap \{y' = 0\}$.

$$\int_\gamma |K| (1-x'^2 + 0(\tau))\, y'dx' =$$

$$\int_{x'_1}^{x'_2} |K| (1-x'^2 + 0(\tau))\, y'_+(x')\, dx' + \int_{x'_1}^{x'_2} |K| (1-x'^2 + 0(\tau))\, |y'_-(x')|\, dx'$$

$$\geq 2.C_2 . \frac{1}{2} .Y\tau^2 .\, 2\tau X = E\, \tau^3 \tag{116}$$

with $E = 2C_2XY > 0$ because of (113), (114), (109) and $x'_1 < -\tau X < \tau X < x'_2$, while for $-\tau X \leq x' \leq \tau X$ we have $|y'_+(x')| \geq \tau^2 Y$ and $|y'_-(x')| \geq \tau^2 Y$ since $\gamma \subset K_0 \backslash R_\tau$.

On the other hand, because of (115) for some $F > 0$:

$$\left| \int_\gamma K\, \tau^{10} (\tilde{\mu}_2 x' + \tilde{\mu}_1)\, dx' \right| \leq F\tau^4 \tag{117}$$

(116) and (117) together show that $\int_\gamma \omega_D$ cannot be zero for τ sufficiently small.

3. Study in a "principal rescaling cone" around the DH-line and the DL-line

We proceed as in B.6, for the saddle case (around the TSC-line). After principal rescaling we come back to the focus-family

$$\bar{y}\frac{\partial}{\partial \bar{x}} + (-\bar{x}^3 + \bar{\mu}_2\bar{x} + \bar{\mu}_1 + \bar{y}(\bar{\nu} + b\bar{x}))\frac{\partial}{\partial \bar{y}} + t(\bar{x}^2 + d\bar{\nu}\bar{x} + O(t))\,\bar{y}\frac{\partial}{\partial \bar{y}} \qquad (1)$$

with $d = \frac{\partial b}{\partial \nu}(0)$.

We work in the principal rescaling charts $\bar{\mu}_2 = \pm 1$, in which the expression is :

$$\bar{y}\frac{\partial}{\partial \bar{x}} + (-\bar{x}^3 \pm \bar{x} + \bar{\mu}_1 + \bar{y}(\bar{\nu} + b\bar{x}))\frac{\partial}{\partial \bar{y}} + t(\bar{x}^2 + d\,\bar{\nu}\,\bar{x} + O(t))\,\bar{y}\frac{\partial}{\partial \bar{y}} \qquad (2)$$

We use the following rescaling :

$$\begin{cases} \bar{x} = x' \\ \bar{y} = y' \end{cases} \qquad \begin{cases} \bar{\nu} = \tau\nu' \\ \bar{\mu}_1 = \tau\mu_1' \\ t = \tau u \end{cases} \qquad (3)$$

changing (2) into :

$$y'\frac{\partial}{\partial x'} + (-x'^3 \pm x' + bx'y')\frac{\partial}{\partial y'} + \tau\,(\mu_1' + \nu'y' + ux'^2y')\frac{\partial}{\partial y'} + O(\tau^2) \qquad (4)$$

An analysis similar as that in B.6 will enable us to prove the following : The bifurcation diagram of (1) intersected with sufficiently small half-spheres $\{\bar{\mu}_1^2 + \bar{\nu}^2 + u^2 = \epsilon^2,\ u \geq 0\}$ around (0,0,0) is as shown in Figure 61.

$\bar{\mu}_2 = +1$ (DL-line) $\bar{\mu}_2 = -1$, (DH-line)

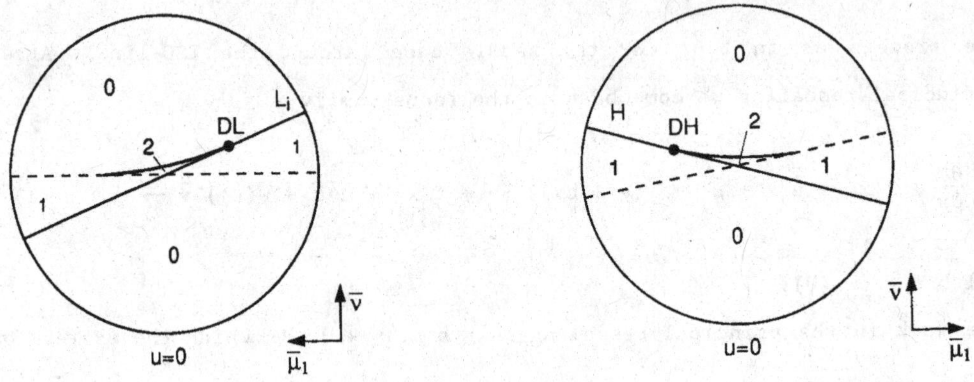

Figure 61

Exactly like in B.5 (compare with the upper part of Figure 54), the dotted line represents the fact that a limit cycle leaves an a priori fixed neighborhood V of (0,0) in $(\bar{x},\bar{y})$-space. For a more detailed description of this phenomenon we refer to B.5.

If we let the boundary ∂V tend to infinity, then the slope of the dotted line tends to 0 (see [Z2]).

CHAPTER VII : CONCLUSIONS AND DISCUSSION OF REMAINING PROBLEMS

In this chapter we will discuss the questions left unsolved after the treatment given in the two preceeding chapters. Some of these questions have already been mentioned in the introduction, but here we want to make them more precise, in terms of a general conjecture and some related subconjectures. Our discussion will concern the lines and points of bifurcation in the diagrams on the 2-sphere S, centered at the origin of the parameter space, as they appear in Figs. 2,3 an 4.

A. The general conjecture

In Chap. V, D.4, we introduced the compact surface of limit cycles defined in $\mathbb{R} \times S$, whose projection in the 2-sphere S of parameters determines the number and nature of limit cycles. In fact, Σ was defined in $\mathbb{R}^2 \times \mathbb{R}$ because of the local chart in which we were working, although it is more natural to regard it as a surface in $\mathbb{R} \times S$. Denote by $\pi: \mathbb{R} \times S \to S$ the projection.

Recall that to each value of u in Σ corresponds a limit cycle Γ_u of X_λ, with $\lambda = \pi(u)$. This limit cycle is hyperbolic if and only if the map π is of maximal rank at u. Therefore the bifurcation set of non hyperbolic (double, or more degenerate) cycles is the critical locus, $C\Sigma$, of π.

General Conjecture. The critical locus $C\Sigma$ is a simple line DC of double cycles, which is a fold line for the projection π.

In the elliptic case, we must add that the relative positions of the lines H, CT_ℓ, DT_s, DC, are as illustrated in Fig. 4, with only one point of transversal intersection of H with CT_ℓ as well as with DT_s, and with one point of transversal intersection of DC with DT_s.

All the unproved assertions about the diagrams proposed in Figs. 2, 3 and 4 reduce to this conjecture. Since this might not be completely obvious, some additional explanations will be given below.

Consider for example the saddle case in Fig. 2. We know that Σ is diffeomorphic to an open disk, and $\bar{\Sigma}$ is a topological disk whose boundary projects onto $H \cup L_\ell \cup L_r$. Our conjecture implies that Σ projects regularly outside the line DC, and the line H cuts $L_\ell \cup L_r$ at a single point (located on L_ℓ, above the point TSC, found by means of the central rescaling). Otherwise the critical locus of Σ would contain some extra portion, distinct from DC. We will see in part B below that the fact that $L_\ell \cup L_r$ cuts H at a single point implies that the relative positions, of H, L_ℓ, L_r, SC_i, SC_s are as illustrated in Fig. 2. Similarly, the conjecture implies that the lines L_ℓ and L_r cannot cut the line H in the focus and elliptic cases, outside the end points TB_r and TB_ℓ.

Notice also that as a consequence of the validity of the conjecture it would follow that the number of limit cycles does not exceed 2. This assertion however seems weaker than the conjecture. Nevertheless it would imply the correctness of the bifurcation diagram in the saddle and in the focus case, without implying the hyperbolicity of the limit cycles outside DC.

Next we present some subconjectures, each of which is related to a particular aspect of the general question concerning limit cycles. The number and nature of these cycles are only established at some parts of the parameter space such as a neighborhood of the points TB and the domain of the central rescaling. Part of the difficulty is related directly to the use of rescalings. Indeed, in the principal or central rescalings, we work on a new phase space $(\bar{x},\bar{y})$ and a new parameter $(\bar{\lambda},r)$, where r is the radial component. Let $\psi:(\bar{x},\bar{y},\bar{\lambda},r)\to(x,y,\lambda)$, be the rescaling map, which decomposes into $\psi_1(\bar{x},\bar{y},\bar{\lambda},r)=(x,y)$, $\psi_2(\bar{\lambda},r) = \lambda$. We have to select a compact domain $\bar{A}$ in the phase space $(\bar{x},\bar{y})$ and some compact domain $\bar{B}$ in the parameter space $\bar{\lambda}$ to obtain a result valid for $r>0$, small enough. The domain in the initial phase space $A_r=\psi_1(\bar{A},r)$ goes to zero when $r\to 0$. Since we have to describe the phase portrait of X_λ in some fixed neighborhood A of the initial (x,y)-

phase space, we must face the problem of extending the picture to the complementary region : $A-A_\tau$, for τ small enough. We will see that this problem is trivial in the saddle case. In the focus case it will be possible to prove it for the central rescaling (see C.2). In the elliptic case we only obtain partial results, essentially in the region with 3 singularities (see D).

B. The saddle case

1. The relative positions of the lines H, L_ℓ, L_r, SC_i, SC_s.

We want to show that this position, as it appears in Fig. 2 is determined by the study of X_λ along the Hopf line H. Recall that in the principal rescaling the family is equivalent to

$$X^P_{\bar\lambda} + O(t),\ \bar\lambda = (\bar\mu_1, \bar\mu_2, \bar\nu) \tag{1}$$

$$X^P_{\bar\lambda} = y\partial/\partial x + [x^3 + \bar\mu_2 x + \bar\mu_1 + y(\bar\nu + bx)]\partial/\partial y \text{ (bars omitted in x,y)} \tag{2}$$

Notice first that the position of the lines H, L_ℓ, etc., for $X^P_{\bar\lambda} + O(t)$ reduces to the same question for $X^P_{\bar\lambda}$ outside a neighborhood W of the point TSC. The reason is that outside W for the family $X^P_{\bar\lambda}$ these lines are stable, since they are defined by a transversality condition.
The study inside sufficiently small neighborhoods of TSC has been made in VI.B.6.

Next, use the following change in parameters and x variable :

$$\bar\mu_1 = (1/27)\ (r-2)(2r^2+r-1),\ \bar\mu_2 = -(r^2-r+1)/3;\ r \in [0,1]$$

$$x = x' + (r-2)/3 \tag{3}$$

Omitting primes, the singular points are located at $x = 0$ (saddle s_1), $x = 1$ (saddle s_2) and $x = 1-r$ (focus e). Using $(r,\bar\nu)$ as new parameter, the internal region I corresponds to $r \in [0,1]$. The saddle node line SN_r corresponds to $r = 0$ and SN_ℓ to $r = 1$. Notice also that the change in parameter is singular at r equal 1 and 0. This is not a real limitation since the bifurcation diagram is already known in a neighborhood of $SN_\ell \cup SN_r$.

The expression of $X^P_{\bar\lambda}$ for the new parameters $\bar\lambda = (r,\bar\nu)$ is as follows :

$$X^P_{\bar\lambda} = y\partial/\partial x + [x(x-1)(x-1+r) + y\ (\bar\nu + b\ (x+(r-2)/3)]\ \partial/\partial y \tag{4}$$

The divergence at (x,0) is given by

$$\operatorname{div} X^P_{\bar\lambda}\ (x,0) = \bar\nu + b\ (x+(r-2)/3) \tag{5}$$

In particular, at e it is given by $\bar\nu$-b/3 (2r-1), and the equation of H becomes

$$(H) \qquad \bar\nu = b/3\ (2r-1) \tag{6}$$

This line passes through the point TSC located at (1/2,0). The remarks made above about the behavior near TSC imply that there exists an $\epsilon > 0$ such that for r at the values $\{1/2 \pm \epsilon\}$ the relative position of the saddle connection and Hopf lines is known. The same is true for r at the values $\{\epsilon, 1-\epsilon\}$. That is, if $\bar\nu = H(r) = b/3\ (2r-1)$, $\bar\nu = SC_i(r)$, etc ..., are equations for each of these lines, then

$$SC_i(r) > H(r) > L_r(r) > SC_s(r), \text{ for } r = \epsilon,\ 1/2 - \epsilon$$

$$SC_s(r) > L_\ell(r) > H(r) > SC_i(r), \text{ for } r = 1-\epsilon,\ 1/2 + \epsilon \tag{7}$$

Now we study the evolution with respect to r of the vector field (4), along the line H. Let $X_r = X^P_{(H(r),r)}$ be this family, depending on the parameter r $\in [0,1]$.

$$X_r = y\frac{\partial}{\partial x} + [x\ (x-1)\ (x-1+r) + y\ b(x+r-1)]\ \frac{\partial}{\partial y}$$

For two values r_0, r_1 of r, calculate

$$< X_{r_0}, X^{\perp}_{r_1} > = y\ [x(x-1)\ (r_1-r_0) + y(\bar{\nu}_1-\bar{\nu}_0) + \ (by/3)\ (r_1-r_0)] \tag{8}$$

for $\bar{\nu}_i = H(r_i) = b/3\ (2r_i-1)$, $i = 0,1$. This gives

$$< X_{r_0}, X^{\perp}_{r_1} > = (r_1-r_0)\ y[x\ (x-1) + by] \tag{9}$$

The sign of this expression, for $r_1 > r_0$ and $x \in [0,1]$ is everywhere positive, except on the region $0 \le y \le x(x-1)/b$.
This implies that the trajectories of X_r have a rotational property with respect to the parameter r, in the region $\{y \le 0; 0 \le x \le 1\}$.

Consider the region $r \in [0,\ 1/2\ -\ \epsilon]$. Because of Chapter V.D (§2) we know that the stable separatrix $W^s(s_1)$, for s_1, is below the unstable one $W^u(s_2)$, for s_2, as shown in Fig. 62, since the line SC_i remains above the line H in the bifurcation diagram.
The same property can also be obtained by using the rotational property with respect to r.

on SC_i on H

Figure 62.

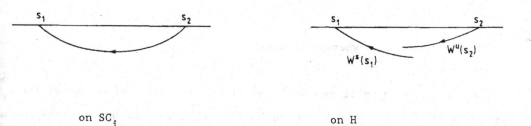

Still for r in $[\epsilon, 1/2 - \epsilon]$, we want to compare the positions of the two separatrices $W^s(s_2)$, $W^u(s_2)$ (on the left of s_2). For each r in $]0,1[$, let $S(r)$ and $U(r)$ denote the intersections of $W^s(s_2)$ and $W^u(s_2)$ with the Ox-axis. The situation, $S(r) > U(r)$, for $r = \epsilon$ and $r = 1/2 - \epsilon$, is illustrated in Fig. 63.

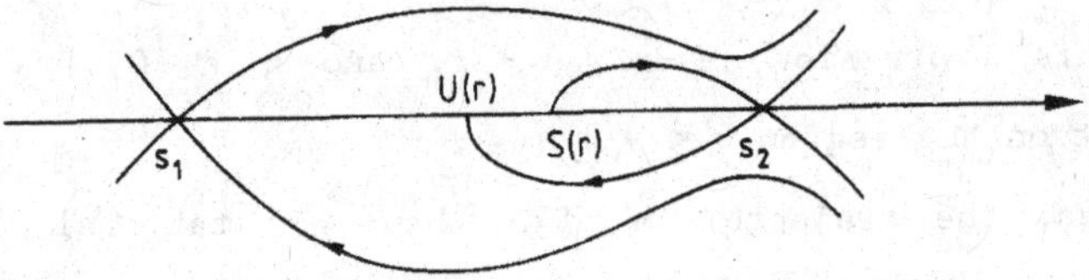

Figure 63.

We formulate the following

Conjecture 1. For every r in $[\epsilon, 1/2 - \epsilon]$, $S(r) > U(r)$.

The validity of this conjecture implies that L_r is entirely below H. Since, for r in $[0, 1/2]$, it is obvious that the line SC_s is below L_r, it follows that the relative positions of the lines SC_s, SC_i, L_r and H are as shown in Fig. 8. A similar conjecture, taking s_1 instead of s_2, can also be formulated for $1/2 < r < 1$.

2. The case where b is small

For b small and any r in $]0,1[$, the family of vector fields $X_r = y\partial/\partial x + [x(x-1)\ (x-1+r) + yb\ (x-(1-r))]\ \partial/\partial y$ can be regarded as a perturbation of the family of Hamiltonian vector fields $X_r^H = y\partial/\partial x - x(1-x)\ (x-1+r)\ \partial/\partial y$. Let H_r be the Hamiltonian function, with $H_r(0) = 0$, whose level curves for r in $]0,1/2[$ are illustrated in Fig. 64

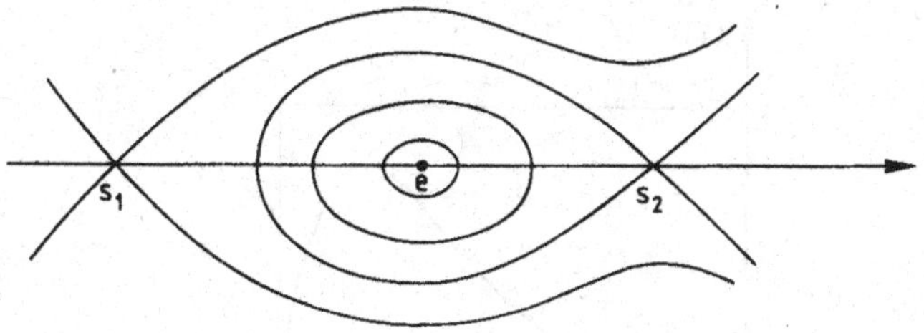

Figure 64.

For r in]0,1/2[, let Γ(r) be the loop at the saddle s_2. From the perturbation theory reviewed in the introduction of Chap. IV, follows that the fact that the line H does not meet L_r. Hence the validity of Conjecture 1 for b small enough, is equivalent to the assertion that :

$$I(r) = \int_{\Gamma(r)} y(x+r-1)\, dx \neq 0 \qquad \text{for any } r \text{ in }]0,1/2[.$$

Similar considerations hold for r in]1/2,1[, integrating on the respective loop at s_1.

A long but elementary calculation of integrals leads to the following formula :

$I(r) = cI_*(r)$, where c is a positive constant and

$$I_*(r) = (1/2-r)\ [1/12(5/2+r)\ (2-r)^2\ \sigma - 3/2\ \sqrt{2}(5/3-r/6+r^2/6)r^{1/2}],$$

with σ defined by $\sigma = \text{Argch}\ [(1+r)/(1-5/2r+r^2)^{1/2}]$.

We see that I_* vanishes for r = 0 and r = 1/2. It is positive on]0,1/2[, has an infinite derivative at 1/2 and $I_*(r) \approx \rho r^{7/2}$ at 0, where $\rho = 0{,}41\ldots$; see Fig. 65. Therefore Conjecture 1 is true for b > 0, small enough.

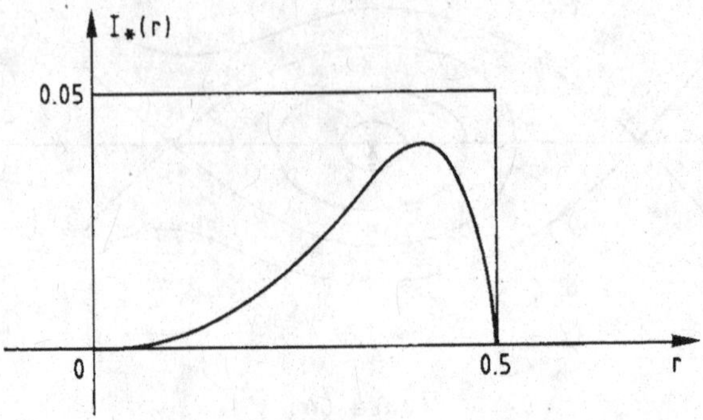

Figure 65.

For $\bar{\nu}$ and b small, the family of vector fields

$$X_{\lambda}^{P} = y\partial/\partial x + [x(x-1)(x-1+r)+y(\bar{\nu}+b(x+1/3(r-2))]\ \partial/\partial y$$

can be considered as a perturbation of the Hamiltonian vector field X_r^H. Using again the perturbation theory for $r \neq 1/2$ (where the Abelian integrals degenerate), we see that for b>0 small enough and outside a neighborhood W around TSC, the bifurcation diagram for the family X_{λ}^{P}, as proposed in Figure 8 (with at most one cycle which grows monotonically with $\bar{\nu}$ for each fixed r) can be reduced to a conjecture about the ratio of elliptic integrals.

In fact, for each $r \neq 1/2$ in $]0,1[$, let $\Gamma(r,h)$ be the cycle for the Hamiltonian H_r around $e=1-r$, on the level $\{H=h\}$; h in $[h_0(r), h_1(r)]$, where $h_0(r) = H_r(1-r)$ and $h_1(r) = H_r(0)$, if $1/2 < r < 1$, or $h_1(r) = H_r(1)$, if $0 < r < 1/2$.

The elliptic integral associated to X_{λ}^{P} is defined by

$$I(r,h)=\int_{\Gamma(r,h)} y[(\bar{\nu}+b(x+1/3(r-2)]\ dx = (\bar{\nu}+b/3(r-2))J_0(r,h) + bJ_1(r,h),$$

where $J_0 = \int_{\Gamma(r,h)} y\ dx$ and $J_1 = \int_{\Gamma(r,h)} y\ xdx$.

Let $P(r,h) = J_1(r,h)/J_0(r,h)$.

Conjecture 2. For each $r \neq 1/2$ in $]0,1[$ fixed, the function P is strictly monotonic and satisfies $P'_h(r,h) \neq 0$ for any h in $[h_0(r),h_1(r)]$.

Remark. This conjecture justifies the bifurcation diagram of X^P_λ as well as the bifurcation diagram of the family X_λ for λ in the sphere S.

3. Passing from A_τ to A

We may choose a neighborhood A and $\bar{A}$ such that the contact along the boundary is as illustrated in Fig. 66.

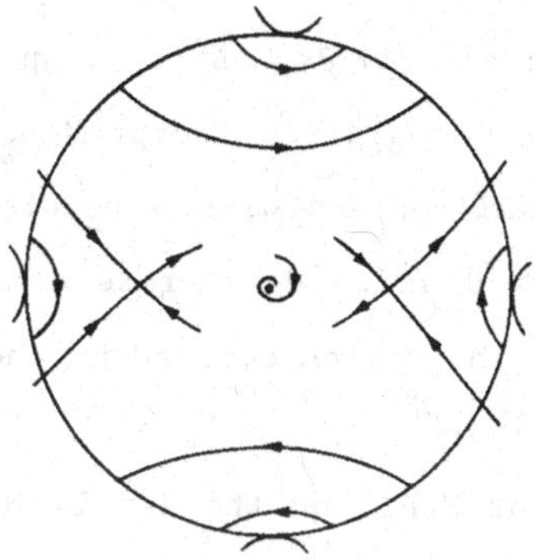

Figure 66.

In the annulus A-A_τ there exist no singular point, and some trajectories go from ∂A to ∂A_τ. This implies, using Poincaré-Bendixson Theorem, that $X_\lambda | A$ is equivalent to $X_\lambda | A_\tau$.

C. The focus case

As we saw in Chapter IV, it is easy to establish the existence and relative position of the lines L_ℓ, L_i, L_r. The relative position of these lines with respect ot the Hopf line H can be reduced to a conjecture similar to Conjecture 1 in the saddle case.

1. **The line DC**

It might be possible to find the entire line DC using the central rescaling. In Chapter VI, we have indicated how to study its end points using this rescaling. Recall the formula obtained for the family, dropping primes in x',y'.

$$(1/\tau)X_\lambda = X'_{\lambda'} = y\partial/\partial x + [(-x^3+\mu'_2x+bxy) + \tau(\mu'_1+\nu'y+yx^2) + yO(\tau^2)]\partial/\partial y$$

The end points of DC were studied taking $\mu'_2 = \pm 1$. It is reasonable to look for the line DC itself outside some small neighborhoods of these points, taking $\mu'_1 = 1$ and μ'_2, ν' as variables. Here we take μ'_1 to be 1 and not -1 because in Fig. 3 the line DC approaches the point DL by the left side.

The family appears as a perturbation of the following symmetric field : $X^S(\mu'_2) = y\partial/\partial x + (-x^3+ \mu'_2x + bxy)\partial/\partial y$. As in Chap VI, it is easy to find a Hamiltonian $H(\mu'_2)$ and an integrating factor $K(\mu'_2)$. We can see it as a one parameter family of Hamiltonians with a regular center for $\mu'_2 < 0$, a lower saddle loop for $\mu'_2 > 0$ and a degenerate center for $\mu'_2 = 0$. $H(\mu'_2)$ is analytic in (x,y,μ'_2) in the region covered by the cycles. See Fig. 67.

Let $\Gamma(\mu'_2,h)$ be the cycle of $H(\mu'_2)$ on the level $\{H(\mu'_2) = h\}$. The related Abelian Integral is

$$J(h,\mu'_2,\nu') = \int_{\Gamma(\mu'_2,h)} K(\mu'_2)\ (1+\nu'y+x^2y)\ dx.$$

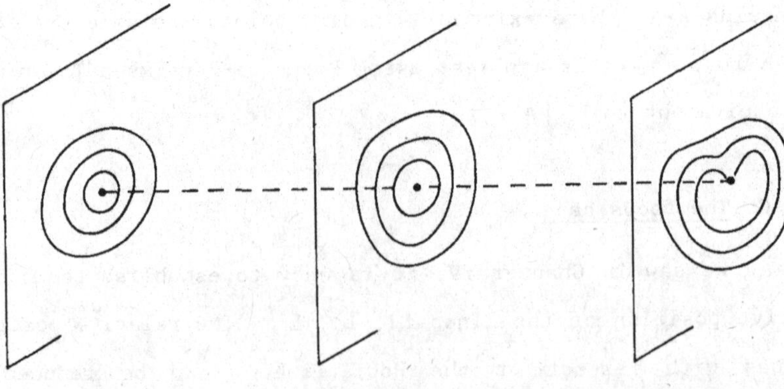

$\mu'_2 < 0$ $\mu'_2 = 0$ $\mu'_2 > 0$

Figure 67.

The equation for the line DC is obtained by eliminating h from

$$J(h,\mu',\nu') = 0 \text{ and } J_h(h,\mu'_2,\nu') = 0.$$

Conjecture 3. These equations define a function $\nu' = \nu'(\mu'_2)$ whose graph is the line DC.

If we write $J = \nu' J_1 + J_2$, where

$J_1 = \int_{\Gamma(\mu'_2,h)} K(\mu'_2)y\, dx$ and $J_2 = \int_{\Gamma(\mu'_2,h)} K(\mu'_2)(1 + x^2y)\, dx$, we have

$\partial J/\partial h = \nu' \partial J_1/\partial h + \partial J_2/\partial h = 0$, which we can solve in ν' if $\partial J_1/\partial h \neq 0$ for all h. Then we can eliminate ν' to find the following equation in h :

$$-[(\partial J_2/\partial h) / (\partial J_1/\partial h)]J_1 + J_2 = 0.$$

Conjecture 3 is equivalent to require that this equation defines a function $h(\mu'_2)$; then the function $\nu'(\mu'_2)$ is given by

$$\nu'(\mu'_2) = -[(\partial J_2/\partial h) / (\partial J_1/\partial h)][h(\mu'_2), \mu'_2].$$

2. Passing from the local analysis in the central rescaling to a fixed domain in phase space

We will prove that in a sufficiently small but fixed neighbourhood V of (0,0) in (x,y)-space (with ∂V transverse to X_λ) all limit cycles can be studied by means of central rescaling. By this we mean that the size of the limit cycles is always schrinking to zero when the parameters approach the origin in the way indicated in the central rescaling : if we take $(\mu_1,\mu_2,\nu) = (r^4\mu'_1, r^2\mu'_2, r^2\nu')$ with $(\mu'_1)^2 + (\mu'_2)^4 + (\nu')^4 = 1$ then for some fixed $C > 0$ and some fixed $\epsilon > 0$ the closed orbits of $X_{(\mu_1,\mu_2,\nu)}$ for $0 < r < \epsilon$ will be contained in some $V_r = \{(x,y) \mid x^4 + y^2 \leq Cr^4\}$. On $V \setminus V_r$

the flow of the vector field $X_{(\mu_1,\mu_2,\nu)}$ will consist of orbits crossing ∂V and having their α-limit set in V_τ.

This is because the case under consideration is the expanding focus. Let us recall its expression :

$$y \frac{\partial}{\partial x} + (-x^3 + \mu_2 x + \mu_1 + y(\nu + b(\lambda)x + x^2 + x^3 h(x,\lambda)) + y^2 Q(x,y,\lambda)) \frac{\partial}{\partial y} \tag{10}$$

Associated to (10) is the family of symmetric 1-forms $\omega^S_{\mu_2}$ with :

$$\omega^S_{\mu_2} = y\,dy + (x^3 - \mu_2 x - b\,x\,y)\,dx \tag{11}$$

The $\omega^S_{\mu_2}$ are symmetric under $(x,y) \to (-x,y)$ and they define the following family of foliations :

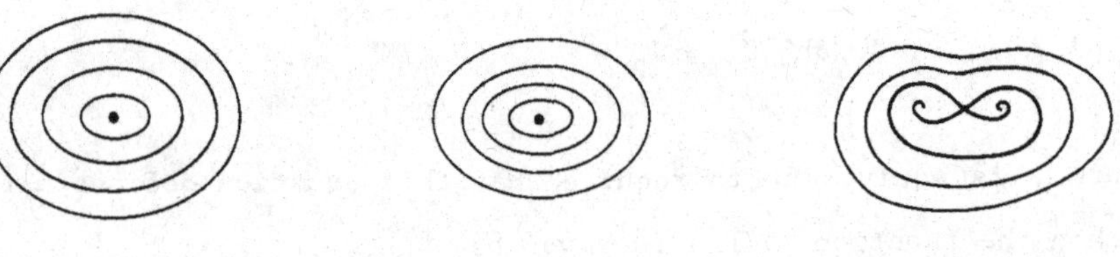

$\mu_2 < 0$ $\mu_2 = 0$ $\mu_2 > 0$

Fig. 68

Let $\phi_\tau : \mathbb{R}^2 \to \mathbb{R}^2$, $(x',y') \to (\tau x', \tau^2 y')$

$$\psi_\tau : \mathbb{R}^3 \to \mathbb{R}^3,\ (\mu_1', \mu_2', \nu') \to (\tau^4 \mu_1', \tau^2 \mu_2', \tau^2 \nu') \tag{12}$$

Let K by any compact neighbourhood of $0 \in \mathbb{R}^2$, bounded by a closed integral curve of ω^S_0. Notice that if $\mu_2 > 0$ is small enough, the set consisting of non-closed curves of $\omega^S_{\mu_2}$ is in the interior of K.

Let A be a compact neighbourhood of $0 \in \mathbb{R}^3$, and consider

$$A_\tau = \psi_\tau(A),\ K_\tau = \phi_\tau(K) \tag{13}$$

we intend to prove the following proposition which will imply the statements made in the introduction of this paragraph.

Proposition

There are neighbourhoods $A°$, $K°$ as above such that for each τ with $0 < \tau \leq 1$ and for each $\lambda \in A^\circ_\tau$, the orbits of X_λ passing through points of $\partial K°$ have for negative time a point in common with ∂K°_τ.

Notice that we do not assert that the orbits are transverse to $\partial K°$ nor to ∂K°_τ. One can however find a fixed $V \subset K°$, V a neighbourhood of 0 in $\mathbb{R}^2$ with ∂V transverse to X_λ; moreover it will follow from the proof of the proposition that the orbits eventually remain in K°_τ when $t \to -\infty$.

Proof

i) The mapping $(\phi_\tau,\psi_\tau) : (x',y',\mu_1',\mu_2',\nu') \to (x,y,\mu_1,\mu_2,\nu)$

$$= (\phi_\tau(x',y'),\ \psi_\tau(\mu_1',\mu_2',\nu'))$$

maps $\tau X'_\tau \,|\, K \times A$ onto $X_\lambda \,|\, K_\tau \times A_\tau$, where

$$X_\tau' = y'\frac{\partial}{\partial x'} + (-x'^3 + \mu_2'x' + bx'y' + \tau(\mu_1' + \nu'y' + y'x'^2) + \tau^2 x'y'k(x',\lambda',\tau) + \tau^2 y'^2 q(x',y',\lambda',\tau))\frac{\partial}{\partial y'} \qquad (14)$$

with $k(x',\lambda',\tau) = \frac{1}{\tau^2}(b(\psi_\tau(\lambda'))-b) + x'^2 h(\tau x',\psi_\tau(\lambda'))$

$$q(x',y',\lambda',\tau) = Q(\phi_\tau(x',y'),\psi_\tau(\lambda'))$$

Fixing any $0 < \tau_0 \leq 1$ we have that $K\backslash K_{\tau_0} = \bigcup_{\tau\in[\tau_0,1]} \partial K_\tau$.

For sufficiently small K and A, we will show that for any τ_0 and any $\lambda \in A_{\tau_0}$ the rays emanating from the origin are transverse to the orbits of X_λ in $K\backslash K_{\tau_0}$ and also that for all points $m \in K\backslash K_{\tau_0}$, the X_λ-orbit through m, for negative times, cuts again the ray $\mathbb{R}^+ m = \{sm \mid s > 0\}$ at some point, rm, with $r < 1$.

ii) To this end we consider $X'_{\tau,\lambda'}(x',y')$ depending on $(\tau,\lambda') \in [0,1] \times \mathbb{R}^3$ and take $M > 0$ such that $|k(x',\lambda',\tau)| \leq M$ and $|q(x',y',\lambda',\tau)| \leq M$ when $(\tau,\mu_1',\mu_2',\nu',x',y') \in [0,1] \times A_1 \times K_1$ for some choice of A_1 and K_1.

To find small $K \subset K_1$ and $A \subset A_1$ with the requested properties we will try to use neighbourhoods of the form $K = \phi_u(K_1)$ and $A = \psi_u(A_2)$ for $u > 0$, where the choice of $A_2 \subset A_1$ as well as the value of u still need to be made precise.

Let $G_u = (\phi_u, \psi_u)$. We have : $\frac{1}{u} G_u^* (X'_{\tau,\lambda'}) = X'^{u}_{\tau,\lambda'}$ where

$$X'^{u}_{\tau,\lambda'}(x",y") = y" \frac{\partial}{\partial x"} + (-x"^3 + \mu_2"x" + bx"y" + \tau u(\mu_1" + \nu"y" + x"^2y") + O(\tau^2u^2)) \frac{\partial}{\partial y"} \tag{15}$$

with X'^u defined on $K_1 \times A_2$.

For u=0, X'^0 is the symmetric vector field $y" \frac{\partial}{\partial x"} + (-x"^3+\mu_2"x"+bx"y") \frac{\partial}{\partial y"}$ tangent to ∂K_1.

So by continuity, for fixed K_1, A_2, $\tau \in [0,1]$, and $0 < u \leq u_1$ with u_1 sufficiently small, $X'^{u}_{\tau,\lambda'}$ is transverse to $\mathbb{R}^+m$ at each $m \in \partial K_1$ and the $X'^{u}_{\tau,\lambda'}$-orbit of m is going to cut $\mathbb{R}^+m$ again for $t < 0$.

As each ϕ_u is linear and $(\phi_u)_*(u X'^{u}_{\tau,\lambda'}) = X'_{\tau,\lambda'}$ we see that for $0 < u \leq u_1$ and $\forall\lambda' \in \psi_u(A_2)$, $X'_{\tau,\lambda'}$ will be transverse to $\mathbb{R}^+m$ at each $m \in \partial(\phi_u(K_1))$; moreover the $X'_{\tau,\lambda'}$-orbit of m is going to cut $\mathbb{R}^+m$ again for $t < 0$.

To check the position of the first return point on $\mathbb{R}^+m$ (for $t < 0$) with respect to m, we are going to make some calculations using an integrating factor $F(x",y",\mu_2')$ for $\omega^S_{\mu_2"}(x",y")$ in some fixed neighbourhood of ∂K_1.

iii) We take A_2 small enough such that for $\lambda" = (\mu_1", \mu_2", \nu") \in A_2$ we have that ∂K_1 lays in the region of closed orbits of $\omega^S_{\mu_2"}$.

F is such that $\omega^S = F\, dH$ for some regular function $H(x",y",\mu_2")$, where $(x",y")$ is in some fixed neighbourhood of ∂K_1. This is possible because for $\mu_2" = 0$, ∂K_1 is a regular level curve of ω^S_o and hence of $H(x",y",0)$.

Now, $dH(X'^{u}_{\tau,\lambda'}) = \frac{1}{F} \omega^S_{\mu_2"}(X'^{u}_{\tau,\lambda'}) =$

$$\tau u \left[\frac{(x"y")^2}{F} + O(\mu_1") + O(\nu") + O(u)\right] \tag{16}$$

Let $\Gamma^u_{\tau,\lambda'}$ be the arc of $X'^u_{\tau,\lambda'}$-orbit between $m \in \partial K_1$ and the first return $\tilde{m}$ on $\mathbb{R}^+ m$ for $t < 0$.

By integration of (16) along $\Gamma^u_{\tau,\lambda'}$ we find :

$$\frac{1}{\tau u}\,[H(m) - H(\tilde{m})] = \int_0^{T^u_m} \frac{(x''y'')^2}{F}\,dt + O(\mu''_1) + O(\nu'') + O(u) \qquad (17)$$

where $T^u_m > 0$ is such that $X'^u_{(\tau,\lambda),T^u_m}(\tilde{m}) = m$; integration is taken along $\Gamma^u_{\tau,\lambda'}$.

When $u \to 0$, the arc $\Gamma^u_{\tau,\lambda'}$ tends uniformly (in a C^∞ sense) to $\Gamma^o = \partial K_1$, which is the corresponding arc for X^o.

But this $\int_0^{T^u_m} \frac{(x''y'')^2}{F}\,dt$ tends to some positive function, depending smoothly on $m \in \partial K_1$.

Let $S(m) = \int_0^{T^o_m} \frac{(x''y'')^2}{F}\,dt$ be this function; integration is now taken along ∂K_1 with the parametrization given by the flow of X^o :

$$\frac{1}{\tau u}\,[H(m) - H(\tilde{m})] = S(m) + O(\mu''_1) + O(\nu'') + O(u) \qquad (18)$$

It now follows that by shrinking A_2 again if necessary (this finally determines A_2), there exists some $0 < u_o \le u_1$ such that $\forall\tau \in\,]0,1]$, $\forall u \in\,]0,u_o]$ and $\forall\lambda'' \in A_2$ we have that $H(\tilde{m}) < H(m)$.

For $X'_{\tau,\lambda'}$ this means that for $0 < u \le u_o$ and $\lambda' \in \psi_u(A_2)$, the negative $X'_{\tau,\lambda'}$-orbit of $m \in \partial(\phi_u(K_1))$ will cut $\mathbb{R}^+ m$ again in some point rm with $r < 1$; u_o does not depend on $\tau \in\,]0,1]$.

iv) We finally define $A^o = \psi_{u_o}(A_2)$ and $K^o = \phi_{u_o}(K_1)$. We choose any $\tau \in\,]0,1]$, take $\lambda \in A^o_\tau = \psi_\tau(A^o)$ and choose $m_1 \in \partial K^o$.

For any $\tau' \in [\tau,1]$, as $\partial K^o_{\tau'} = \partial(\phi_{\tau' u_o}(K_1))$ and $\lambda \in \psi_\tau(A^o) \subset \psi_{\tau'}(A^o) = \psi_{\tau' u_o}(A_2)$ we know that the negative $X_{\tau',\lambda'}$-orbit (with $\psi_{\tau'}(\lambda') = \lambda$) of $m_{\tau'}$, with $(m_{\tau'}) = \mathbb{R}^+ m_1 \cap \partial K^o_{\tau'}$, will cut $\mathbb{R}^+ m_1$ again in a point $m'_{\tau''} = r'_{\tau'} m'_{\tau'}$ with $r'_{\tau'} < 1$.

Since $\psi_{\tau'}(\lambda')=\lambda$ and $(\phi_{\tau'})_*(\tau' X'_{\tau',\lambda'}) = X_\lambda$ this shows that $\forall\ m_{\tau'}$ the negative X_λ-orbit of $m_{\tau'}$ will cut $\mathbb{R}^+ m_1$ again in a point $m_{\tau''} = r_{\tau'} m_{\tau'}$ with $r_{\tau'} < 1$ and hence $\tau'' < \tau'$. As $[\tau,1]$ is compact this shows that the negative X_λ-orbit of m_1 will finally have to cut ∂K^o_τ.

Using the same argument for $0 < \bar{\tau} < \tau$ sufficiently small will not only show that for $\lambda \in A^o_{\bar{\tau}}$ the X_λ-orbit of a point $m_1 \in \partial K^o$ is going to cut ∂K^o_τ for negative times, but is also going to have its α-limit set in K^o_τ.

Remark

Since the proposition merely gives the existence of domains A^o, K^o with the required properties, in order to gain information on a neighbourhood in (x,y)-space by using central rescaling, it is essential that in the (x',y')-space we work on a sufficiently big compact neighbourhood of 0.
The reason of this need is that when working with central rescaling one often considers parameters in a sector S in some neighbourhood B^o with $S \cap \partial B^o \neq \emptyset$ and $B^o \supset A^o$ and exactly the parameters in $S \cap \partial B^o$ are important. (like f.i. $\mu'_2 = \pm 1$ and $(\mu'_1,\nu') \in K$) In that case we need to work on some $L^o = \phi_r(K^o)$ with r such that $\psi_r(A^o) \supset B^o$ in order to use the proposition in passing from $\phi_\tau(L^o)$ to K^o when $\lambda \in \psi_\tau(S) \subset \psi_\tau(B^o)$.
Indeed if for parameter values λ in $\psi_\tau(B^o)$ we have a complete knowledge of X_λ inside $\phi_\tau(L^o)$, then as $\psi_\tau(B^o) \subset \psi_\tau(\psi_r(A^o)) = \psi_{\tau r}(A^o)$ and $\phi_\tau(L^o) = \phi_\tau(\phi_r(K^o)) = \phi_{\tau r}(K^o)$ we can pass from $\phi_\tau(L^o) = \phi_{\tau r}(K^o)$ to K^o by using the proposition.

D. The elliptic case

We recall the expression of the elliptic family

$$y\frac{\partial}{\partial x} + (-x^3+\mu_2 x+\mu_1+y(\nu+b(\lambda)x + x^2 + x^3 h(x,\lambda)) +y^2\ Q(x,y,\lambda))\frac{\partial}{\partial y}\ , \tag{19}$$

where $b(0) = b > 2\sqrt{2}$.

Concerning (19) we are going to investigate the following facts, which will be valid in a fixed neighbourhood V of (0,0) in the (x,y)-coordinate plane and in a fixed neighbourhood of (0,0,0) in the parameter space (μ_1, μ_2, ν).

i) For $\mu_1 = \mu_2 = 0$ and $\nu > 0$ (resp. $\nu < 0$), the phase portrait of $X_{(0,0,\nu)}$ is like in figure 69 (resp. figure 70),

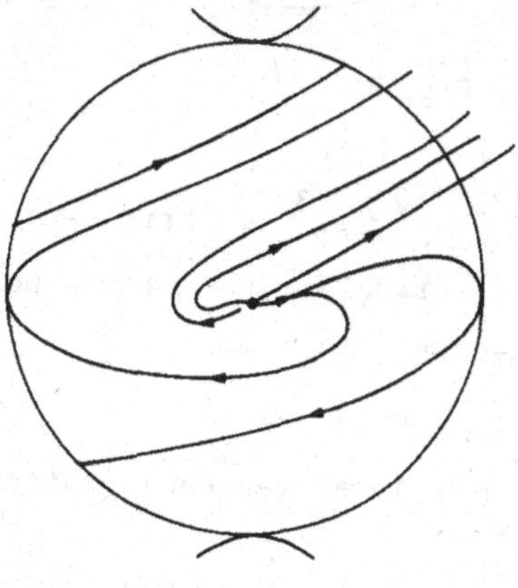

Fig. 69

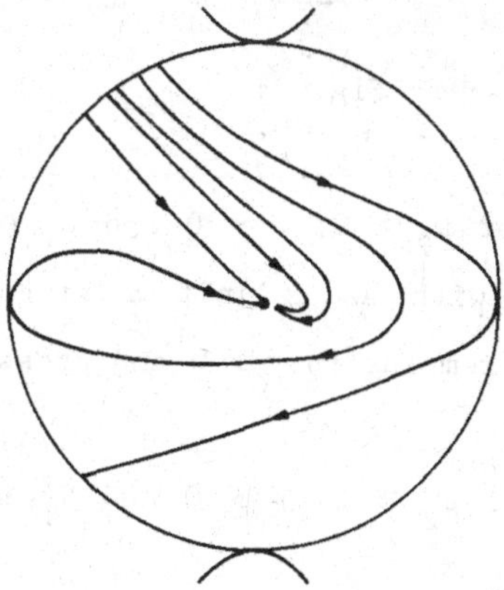

Fig. 70

ii) For $\mu_2 \geq 0$, $\nu \geq 0$ and $\mu_1 \in [\, -\frac{2}{3\sqrt{3}} \mu_2^{3/2}, \frac{2}{3\sqrt{3}} \mu_2^{3/2} [$ there exists an orbit whose α-limit is a singularity (a saddle or a saddle-node) which remains in $y > 0$ and crosses the boundary of V.

iii) For $\mu_2 \geq 0$, $\nu \geq 0$ and $\mu_1 \geq \frac{2}{3\sqrt{3}} \mu_2^{3/2}$ the inner tangency orbits are like in figure 71 and for $\mu_1 = \frac{2}{3\sqrt{3}} \mu_2^{3/2}$ there exists an orbit having as α-limit a singularity (a saddle-node or an nilpotent cusp-point), which remains in $y > 0$ for t increasing and crosses the boundary of V.

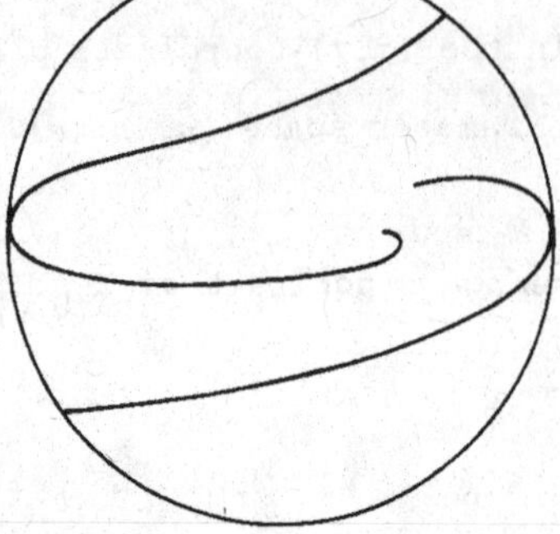

Fig. 71

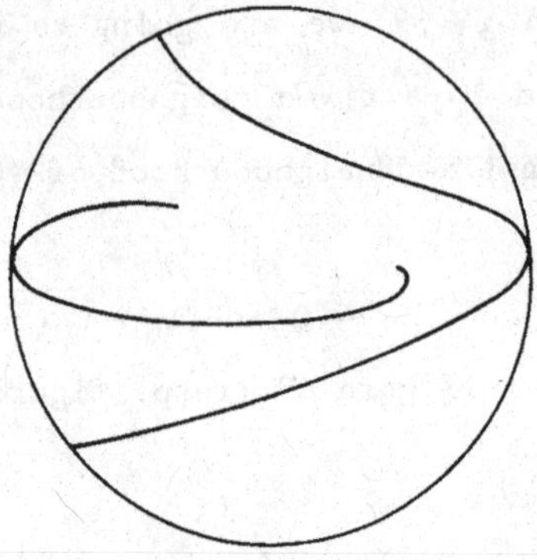

Fig. 72

iv) For $\mu_2 \geq 0$, $\nu \leq 0$ and $\mu_1 \in]-\frac{2}{3\sqrt{3}}\mu_2^{3/2}, \frac{2}{3\sqrt{3}}\mu_2^{3/2}]$ there exists an orbit with as ω-limit a singularity (resp. a saddle or a saddle node), which remains in $y > 0$ and crosses the boundary of V.

v) For $\mu_2 \geq 0$, $\nu \leq 0$ and $\mu_1 \leq -\frac{2}{3\sqrt{3}}\mu_2^{3/2}$ the inner tangency orbits are like in figure 72 and for $\mu_1 = -\frac{2}{3\sqrt{3}}\mu_2^{3/2}$ there exists an orbit having as ω-limit a singularity (a saddle node or a nilpotent cusp-point) which remains in $y > 0$ for t decreasing and crosses the boundary of V.

Let us now verify the assertions made in i) to v).

1. Region $\nu \geq 0$, $\mu_2 \geq 0$, $\mu_1 \geq -\frac{2}{3\sqrt{3}}\mu_2^{3/2}$

Along the curve $y = \frac{\alpha}{2}(x^2 - \frac{\mu_2}{3})$, for $x > (\frac{\mu_2}{3})^{1/2}$, (20)

with $\alpha > 0$ and $\alpha^2 - b\alpha + 2 = 0$, we will calculate the direction of the family $X_{(\mu_1,\mu_2,\nu)}$, nl.

$$\frac{\partial}{\partial x} + [\frac{1}{y}(-x^3+\mu_2 x+\mu_1) + \nu + b(\lambda)x + x^2 + x^3 h(x,\lambda) + yQ(x,y,\lambda)]\frac{\partial}{\partial y} \quad (21)$$

and compare it with the tangential direction of (20) :

$$\frac{\partial}{\partial x} + \alpha x \frac{\partial}{\partial y} \qquad (22)$$

We will prove that $<(21)-(22), \frac{\partial}{\partial y}> \geq 0$ on (20).

Let us write $\mu_1 = -\frac{2}{3\sqrt{3}} \mu_2^{3/2} + \bar{\mu}_1$ with $\bar{\mu}_1 \geq 0$ (23)

and $b(\lambda) = b + \mu_1\theta(\mu_1) + \mu_2\psi(\mu_1,\mu_2) + \nu\phi(\mu_1,\mu_2,\nu)$ (24)

where θ, ψ and ϕ are C^∞.

Now $<(21)-(22), \frac{\partial}{\partial y}> =$

$$\frac{1}{\frac{\alpha}{2}(x^2-\frac{\mu_2}{3})}\left(-x\,(x^2-\frac{\mu_2}{3}) + \frac{2}{3}\mu_2\,(x-(\frac{\mu_2}{3})^{1/2})\right) + bx + \frac{\bar{\mu}_1}{\frac{\alpha}{2}(x^2-\frac{\mu_2}{3})} + \nu + (b(\lambda)-b)x$$

$$+ x^2 + x^3h(x,\lambda) + \frac{\alpha}{2}(x^2-\frac{\mu_2}{3})\,Q(x, \frac{\alpha}{2}(x^2-\frac{\mu_2}{3}),\lambda) - \alpha x$$

$$= (-\frac{2}{\alpha} + b - \alpha)\,x + \frac{4}{3}\,\frac{\mu_2}{\alpha(x^2+(\frac{\mu_2}{3})^{1/2})} + \frac{2\bar{\mu}_1}{\alpha\,(x^2-\frac{\mu_2}{3})} + \nu + (b(\lambda)-b)x + \frac{\mu_2}{3}$$

$$+ (x^2-\frac{\mu_2}{3}) + (x^2-\frac{\mu_2}{3})xh(x,\lambda) + \frac{\mu_2}{3}\,xh(x,\lambda) + \frac{\alpha}{2}(x^2-\frac{\mu_2}{3})\,Q(x, \frac{\alpha}{2}(x^2-\frac{\mu_2}{3}),\lambda)$$

$$= \mu_2\left(\frac{1}{3} + \frac{4}{3\alpha(x+\frac{\mu_2}{\sqrt{3}})} + \frac{1}{3}\,xh(x,\lambda) + x\psi(\mu_1,\mu_2) - \frac{2}{3\sqrt{3}}\,\mu_2^{1/2}\,\theta(\mu_1)x\right)$$

$$+ \bar{\mu}_1\left(\frac{2}{\alpha(x^2-\frac{\mu_2}{3})} + x\theta(\mu_1)\right) + \nu(1 + x\phi(\mu_1,\mu_2,\nu))$$

$$+ (x^2-\frac{\mu_2}{3})\left(1 + xh(x,\lambda) + \frac{\alpha}{2}\,Q(x, \frac{\alpha}{2}(x^2-\frac{\mu_2}{3}),\lambda)\right)$$

It is clear that this expression is ≥ 0 if we take (x,y,μ_1,μ_2,ν) (in a uniform way) sufficiently small.

For $-\frac{2}{3\sqrt{3}}\mu_2^{3/2} \le \mu_1 \le \frac{2}{3\sqrt{3}}\mu_1^{3/2}$ this gives the situation described in figure 73 (for $\mu_2 > 0$)

Besides the position of the direction of the vector field (21) with respect to that of the parabola $\frac{\alpha}{2}(x^2 - \frac{\mu_2}{3})$ we also know that (21) is pointing upwards along $[s,(\frac{\mu_2}{3})^{1/2}] \times \{0\}$ where s denotes the middle singularity (resp. the saddle-node or cusp-like singularity when $\mu_1 = \frac{2}{3\sqrt{3}}\mu_2^{3/2}$).

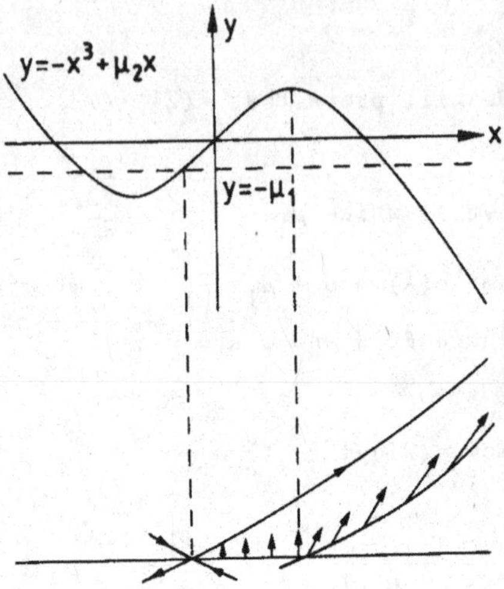

Figure 73

For $\mu_1 > \frac{2}{3\sqrt{3}}\mu_3^{3/2}$ there is no singularity on $]-\infty, (\frac{\mu_2}{3})^{1/2}[\times \{0\}$ and the vector field is everywhere pointing upwards there.

This surely implies i) and iii), and also ii) when $\mu_1 \neq -\frac{2}{3\sqrt{3}}\mu_2^{3/2}$.
For $\mu_1 = -\frac{2}{3\sqrt{3}}\mu_2^{3/2}$ (when s is on the parabola) we need to calculate the slope of the unstable manifold in $s = ((\frac{\mu_2}{3})^{1/2}, 0)$ and compare it with $\alpha(\frac{\mu_2}{3})^{1/2}$.

The 1-jet of the vector field at $((\frac{\mu_1}{3})^{1/2}, 0)$ is

$$\begin{pmatrix} 0 & 1 \\ 0 & \nu + b(\lambda)(\frac{\mu_2}{3})^{1/2} + \frac{\mu_2}{3} + \frac{\mu_2^{3/2}}{3\sqrt{3}} h((\frac{\mu_2}{3})^{1/2}, \lambda) \end{pmatrix},$$

where $\lambda = (-\frac{2}{3\sqrt{3}}\mu_2^{3/2}, \mu_2, \nu)$.

As the slope of the expanding eigenvalue $(\mu_1^2 + \mu_2^2 + \nu^2 > 0)$ is

$\nu + b(\lambda)(\frac{\mu_2}{3})^{1/2} + \frac{\mu_2}{3} + \frac{\mu_2^{3/2}}{3\sqrt{3}} h((\frac{\mu_2}{3})^{1/2}, \lambda)$, we see that it is going to be bigger than $\alpha(\frac{\mu_2}{3})^{1/2}$ for parameter values which are sufficiently small since $\alpha < b$.

Notice that for $\lambda=0$, $y = \frac{\alpha}{2} x^2$ (with $\alpha^2 - b\alpha + 2 = 0$) is a solution curve of $ydy - (-x^3+bxy)\,dx = 0$ and the slope of X_o along $y = \frac{\alpha}{2} x^2$ is also bigger than αx for $x > 0$.

2. Region $\nu \le 0$, $\mu_2 \ge 0$, $\mu_1 \le \frac{2}{3\sqrt{3}} \mu_2^{3/2}$

We prefer to change (19) by means of $(x,y,t) \to (-x,y,-t)$ into

$$y \frac{\partial}{\partial x} + (-x^3 + \mu_2 x - \mu_1 + y(-\nu + b(\lambda)x - x^2 + x^3 h(x,\lambda) - yQ(-x,y,\lambda))) \frac{\partial}{\partial y} \qquad (25)$$

The presence of the minus sign in front of $x^2 y \frac{\partial}{\partial y}$ prevents us from using the same curve in order to prove the similar result obtained in the case $\nu \ge 0$.

We again write $\mu_1 = \frac{2}{3\sqrt{3}} \mu_2^{3/2} - \bar{\mu}_1$ with $\bar{\mu}_1 \ge 0$, changing (25) into

$$y \frac{\partial}{\partial x} + (-x(x^2 - \frac{\mu_2}{3}) + \frac{2}{3} \mu_2 (x - (\frac{\mu_2}{3})^{1/2}) + bxy + \bar{\mu}_1 + y(-\nu + (b(\lambda) - b)x - x^2 + x^3 h(x,\lambda)$$

$$- yQ(-x,y,\lambda))) \frac{\partial}{\partial y} \qquad (26)$$

Now choose any $2\sqrt{2} < b' < b$. (27)

We are going to work along the curve

$$y = \frac{\alpha}{2} (x^2 - \frac{\mu_2}{3}) \quad \text{for } x > (\frac{\mu_2}{3})^{1/2} \qquad (28)$$

with $\alpha > 0$ and $\alpha^2 - b'\alpha + 2 = 0$.

The slope of (28) is αx and we need to compare it to the slope of the vector field along (28).

We want to show that the following expression is nonnegative when $x > (\frac{\mu_2}{3})^{1/2}$:

$$(-\frac{2}{\alpha} + b'-\alpha)x + \frac{4}{3}\frac{\mu_2}{\alpha(x + (\frac{\mu_2}{3})^{1/2})} + \frac{2\bar{\mu}_1}{\alpha(x^2 - \frac{\mu_2}{3})} + (b- b')x - \nu + (b(\lambda)-b)x$$

$$- x^2 + x^3 h(-x,\lambda) - \frac{\alpha}{2}(x^2 - \frac{\mu_2}{3})\, Q(x, \frac{\alpha}{2}(x^2 - \frac{\mu_2}{3}),\lambda)$$

$$= \frac{4}{3}\frac{\mu_2}{\alpha(x + (\frac{\mu_2}{3})^{1/2})} + \frac{2\bar{\mu}_1}{\alpha(x^2 - \frac{\mu_2}{3})} - \nu + x((b-b') + (b(\lambda)-b) - x + x^2 h(-x,\lambda)$$

$$- \frac{\alpha}{2x}(x^2 - \frac{\mu_2}{3})\, Q(x, \frac{\alpha}{2}(x^2 - \frac{\mu_2}{3}),\lambda))$$

This expression is certainly nonnegative for $x > (\frac{\mu_2}{3})^{1/2}$ if we take (x,y) and (μ_1,μ_2,ν) sufficiently small.
Exactly like in the case $(\nu \geq 0)$ this implies the statements in iv) and v).

As a consequence of the assertions made in i) to v), and using also the divergence of (1) where necessary, we obtain the following results :

- There are no closed orbits surrounding more than one singularity.
- Along the surface S of parameter values where the div X_λ is zero in the saddle point, we have the phase portrait as given in fig. 74, containing no closed orbits.

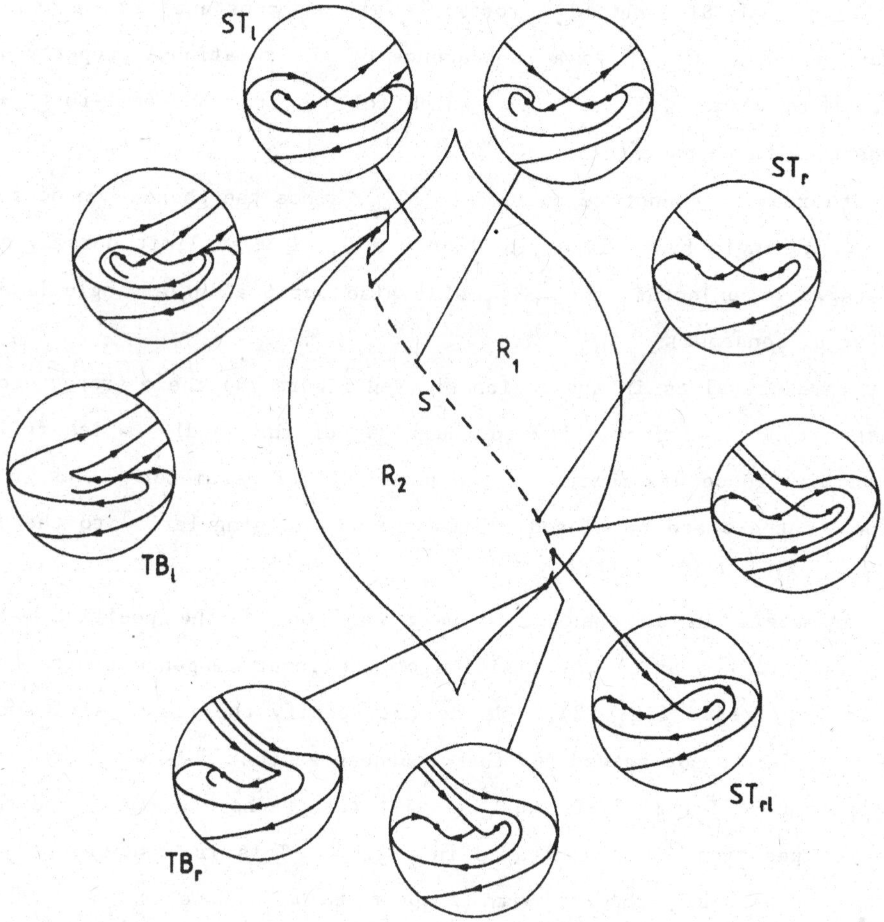

Figure 74

This result will follow from a), b), c) below.

a) The existence of an orbit going from the boundary to a singularity.

b) The fact that the divergence equated to zero separates V (supposed sufficiently small) in two half planes, where the divergence has constant sign, and whose common boundary passes through the saddle (resp. nilpotent cusp point).

c) The fact that ST_ℓ and ST_r (restricted to μ_2 constant) are graphs of functions of μ_1 (which is a consequence of the rotational property with respect to ν) as well as of ν (which follows from the semi-rotational property with respect to μ_1).

This last argument cannot be applied to $ST_{r\ell}$, since the phenomenon does not happen in a single half plane $y \le 0$ or $y \ge 0$, we must limit ourselves to formulate as a conjecture that $ST_{r\ell}$ will also cut S along a single line (a point for μ_2 constant).

- For parameter values in the region R_1 (see figure 74) there are no closed orbits around the singularity to the right of the saddle, which follows from a divergence argument. In the same way for parameter values in the region R_2 there are no closed orbits around the singularity to the left of the saddle.
- For parameters in an open conic neighbourhood of the positive ν-axis (resp. negative ν-axis) the position of the inner tangency orbits is as in fig. 71 (resp. fig. 72). We do not specify the flatness of such a cone but this is not needed for the consequences that this will imply.
- The lines L_ℓ, ST_ℓ, ST_r, $ST_{1\ell}$ and L_r exist (as graphs $\nu = \nu(\mu_1)$) and their relative position is as indicated in fig. 4. This is essentially based on the rotational property with respect to ν. However the relative position of these lines with respect to the line of Hopf-bifurcation is conjectural.
- The lines DT and CT exist (as graphs $\nu = \nu(\mu_1)$) and their relative position is as indicated in fig. 4. We however cannot yet study the relative position of these lines with respect to the Hopf line H, neither the transversal properties (like hyperbolicity, etc.) of the closed orbits, which touch the boundary of V when a parameter value is in CT.

Besides the assertions made in i) → v) and their consequences just made, there is another important fact which can be proven :

vi) All closed orbits for the elliptic family (19) need to pass through a sufficiently large central rescaling chart, i.e. through

$R_\tau = \{(x,y) \mid |x| \le K\tau, |y| \le K'\tau^2\}$ for some $K > 1$ and some $K' > 0$, when we take $(\nu,\mu_1,\mu_2) = (\nu'\tau^2,\mu_1'\tau^4,\mu_2'\tau^2)$ and let $\tau \to 0$.

This implies that even the closed orbits which touch the boundary of an a priori given neighbourhood of (0,0) need to pass through that central rescaling chart, so that this phenomenon of cycle tangency (CT) can presumably be studied using central rescaling.

The only restriction on $K > 1$ is that R_τ should contain all the singularities of X_λ for $\lambda = (\nu'\tau^2, \mu_1'\tau^4, \mu_2'\tau^2)$ and $\max(|\nu'|, |\mu_1'|, |\mu_2'|) \le 1$. K' will be made precise at the end.
Let us from now on impose these restrictions on (ν', μ_1', μ_2') and K. Let us also restrict (x,y) to some compact $A_0 \supset R_\tau$ as in I.2.C.

The proof of assertion vi) is quite similar to that of the other ones.

We consider the parabola

$$y = \frac{\alpha}{2}(x^2 - \mu_2) \tag{29}$$

with $\alpha > 0$ and $\alpha^2 - b'\alpha + 2 = 0$ for some $2\sqrt{2} < b' < b$.

We restrict to points $(x,y) \in A_0 \backslash R_\tau$ and want to prove that the direction vector of X_λ along the parabola are as in figure 75. We also draw X_λ along the x-axis and along part of the boundary.

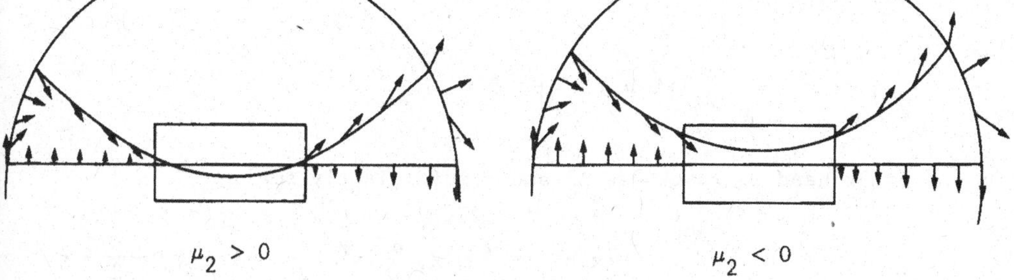

Figure 75.

Using expression (19) we are first going to show that for (x,y) in A_o on the parabola with $x > K\tau$, we have that the following expression is positive :

$$\frac{1}{\frac{\alpha}{2}(x^2-\mu_2)} (-x^3+\mu_2 x+\mu_1) + (\nu+b(\lambda)x + x^2 + x^3 h(x,\lambda))$$

$$+ \frac{\alpha}{2}(x^2-\mu_2)\, Q(x, \frac{\alpha}{2}(x^2-\mu_2),\lambda) - \alpha x$$

$$= -\frac{2}{\alpha}x + b'x - \alpha x + \frac{\mu_1}{\frac{\alpha}{2}(x^2-\mu_2)} + \nu$$

$$+ x\,[(b-b') + (b(\lambda)-b) + x + x^2 h(x,\lambda) + \frac{x\alpha}{2} Q(x, \frac{\alpha}{2}(x^2-\mu_2),\lambda)]$$

$$- \frac{\alpha\mu_2}{2} Q(x, \frac{\alpha}{2}(x^2-\mu_2),\lambda) \qquad (30)$$

In (30) we have

$$-\frac{2}{\alpha}x + b'x - \alpha x = \frac{x}{\alpha}(-2 + b'\alpha - \alpha^2) = 0$$

As $x^2-\mu_2 \geq (K^2-\mu_2')\,\tau^2 \geq (K^2-1)\,\tau^2$, there exists some $C_1 > 0$ with

$$\left|\frac{\mu_1}{\frac{\alpha}{2}(x^2-\mu_2)}\right| \leq C_1\tau^2 .$$

There also exist $C_2 > 0$ and $C_3 > 0$ with

$$|\nu| \leq C_2\tau^2$$

$$\left| -\frac{\alpha\mu_2}{2} Q(x, \frac{\alpha}{2}(x^2-\mu_2),\lambda)\right| \leq C_3\tau^2$$

On the other hand if we choose A_o and τ sufficiently small

$$x[(b-b') + (b(\lambda)-b) + x + x^2 h(x,h) + \frac{x\alpha}{2} Q(x, \frac{\alpha}{2}(x^2-\mu_2),\lambda)] \geq K\frac{(b-b')}{2}\tau$$

implying that (30) will be positive for τ sufficiently small.

A same argumentation can be made for $x < -K\tau$ by using expression (25) instead of (19). (25) is obtained from (19) by means of $(x,y,t) \to (-x,y,-t)$.

To finish the proof we observe that for $|x| = \tau K$, we have

$$y = \frac{\alpha}{2}(K^2 - \mu_2')\tau^2 ;$$

if we take $K' = \frac{\alpha}{2}(K^2 + 1)$ then the parabola (29) will cut ∂R_τ as indicated in fig. 75.

<u>REFERENCES</u>

[A] : V. Arnol'd

Chapitres supplémentaires de la théorie des equations différentielles ordinaires, Ed. Mir, Moscow, 1980

[A2] : V. Arnol'd

Lectures on bifurcations in versal families, Russian Math. Surveys V, 26, 1971

[A.L.] : A. Andronov, E. Leontovich, et al.

Theory of bifurcations of Dynamical Systems on a Plane, I.P.S.T., Jerusalem, 1971

[B.K.K.] : A.D. Basikin, Yu Kuznietzov, A.I. Khibnik

Bifurcational diagrams of dynamical systems on the plane, Computer Center Acad. Sciences URSS, Puschino, 1985

[B] : R. Bogdanov

- Versal deformations of a singular point of a vector field on the plane in the case of zero eigenvalues. (R) Seminar Petrovski, 1976, (E) Selecta Mathematica Sovietica, vol. 1, 4, 1981, 389-421.
- Bifurcation of a limit cycle for a family of vector fields on the plane, (R) Seminar Petrovski, 1976, (E) Selecta Math. Sov., vol.1, 4, 1981, 373-388.

[D] : F. Dumortier

- Singularities of vector fields on the plane, J. Diff. Equat., vol 23, 1 (1977), 53-106
- Singularities of vector fields. Monografias de Matemática 32 IMPA, Rio de Janeiro, 1978

[D.G.] : G. Dangelmayer, J. Guckenheimer

On a four parameter family of planar vector fields, Arch. Rat. Mech. Anal., 97, 1987, 321-352.

[D.R.] : F. Dumortier, C. Rousseau
Cubic Liénard equations with linear damping, Nonlinearity, to appear.

[D.R.S.] : F. Dumortier, R. Roussarie, J. Sotomayor
Generic 3-parameter families of vector fields on the plane, unfolding a singularity with nilpotent linear part. The cusp case. Ergodic theory and dynam. systems, 7, 1987, 375-413

[G.H.] : J. Guckenheimer, P. Holmes
Non-linear oscillations, dynamical systems, and bifurcations of vector fields, Appl. Math. Sc. 42, Springer-Verlag, 1983

[H.C.] : J.K. Hale, S.-N. Chow
Methods of bifurcation theory, Springer-Verlag, Berlin, 1982

[R] : R. Roussarie
On the number of limit cycles which appear by perturbation of separatrix loop of planar vector fields, Bol. Soc. Bras. Mat., Vol. 17, 2, 1986, 67-101

[Sc] : S. Schecter
The Saddle-node separatrix-loop bifurcation, SIAM Journ. Math. Anal., Vol. 18, 4, 1987, 1142-56.

[Se] : A. Seidenberg
A New decision method for elementary algebra, Ann. of Math. 60, 1954, 365-374

[S] : J. Sotomayor
Generic one-parameter families of vector fields on two-dimensional manifolds, Publ. Math. I.H.E.S., Vol. 43, 1974

[S2] : J. Sotomayor
Curvas definidas por equações diferenciais no plano, IMPA, Rio de Janeiro, 1981

[St] : D. Stowe

Linearization in two dimensions, Journ. of Diff. Equat. 63, 1986, 183-226

[T] : F. Takens

Unfoldings of certain singularities of vector fields. Generalized Hopf bifurcations. Journ. of Diff. Equat. 14, 1973, 476-493

[T2] : F. Takens

Forced oscillations and bifurcations.
In : Applications of Global Analysis I, Communications of Math. Inst. Rijksuniv. Utrecht, 3, 1974

[Te] : M.A. Teixeira

Generic bifurcation in manifolds with boundary, J. Diff. Equat., vol. 25, 1, 65-89, 1977

[Z1] : H. Zoladek

Abelian integrals in unfoldings of cod. 3 singular planar vector fields, Part II. The saddle and elliptic case. This volume

[Z2] : H. Zoladek

Abelian integrals in unfoldings of cod. 3 singular planar vector fields, Part III. The focus case. This volume

ABELIAN INTEGRALS IN UNFOLDINGS OF CODIMENSION 3 SINGULAR PLANAR VECTOR FIELDS

PART I. THE WEAKENED 16-TH HILBERT PROBLEM

PART II. THE SADDLE AND ELLIPTIC CASES

PART III. THE FOCUS CASE

Abstract

In this work it is shown that, for small β_i, the system $\dot{x} = y$, $\dot{y} = \pm x + \alpha x^3 + xy + \beta_0 + \beta_1 y + \beta_2 x^2 y$ has at most two limit cycles when $\alpha \in (-1/8, \infty)\setminus\{0\}$ (Part II) and also when $\alpha < -1/8$ (Part III). Part I contains an introduction to the problem, applications of Abelian integrals and some general results.

Table of contents

PART I. THE WEAKENED 16-TH HILBERT PROBLEM

PART II. THE SADDLE AND ELLIPTIC CASES

PART III. THE FOCUS CASE

PART I. THE WEAKENED 16-TH HILBERT PROBLEM

1. Formulation of the problem

In [**1**] Arnold stated the following problem: find the number of zeroes of the integral

$$I(h) = I_\omega(h) = \int \omega, \quad \omega = P(x,y)dx + Q(x,y)dy \tag{1}$$

along the ovals of the curve $H(x,y) = h$, where P, Q and H are polynomials.

This problem comes from the investigation of limit cycles in a polynomial perturbation of a Hamiltonian system on the plane. This connects the integral (1) with 16-th Hilbert Problem.

One can also formulate another weakened version of 16-th Hilbert Problem. Consider a polynomial planar vector field, not necessarily Hamiltonian, with a first integral H, and a small polynomial perturbation. Then the problem of limit cycles in the perturbed system leads to the investigation of Abelian integrals (1) with rational functions P and Q and often non-algebraic curves $H(x,y) = h$. Throughout this paper we consider both problems as the Weakened 16-th Hilbert Problem.

The investigation of zeroes of Abelian integrals is a rather young theory. There are already various (not many) results but each proof uses a different method and no general approach has been developed. However in some cases the situation became clear after prolongation to the complex domain and using the apparatus of complex algebraic geometry (disappearing cycles, monodromy etc.)

2. Finiteness properties

The first general results about zeroes of the function $I(h)$ with P, Q and H-polynomials are due to Khovansky [**16**] and Varchenko [**25**]. They showed that the number of zeroes is bounded by a constant depending on the degrees of P, Q and H. No concrete bound is given.

For the general Weakened 16-th Hilbert Problem there is only the result, announced by Martinet, Moussou, Ramis, Ecalle [**12**] and Iliashenko [**15**], that any planar vector field has finite number of limit cycles. The assertion is about an individual vector field and no uniform or local estimatives are known.

3. The Petrov's bounds

In the study of our problem the following definition is natural.

DEFINITION 3.1. *A space of functions on the domain Ω is called Chebyshev with accuracy k (in Ω) if the number of zeroes of every non-zero function from this space is less than its dimension plus k. The spaces with accuracy 0 are called Chebyshev.*

Petrov considered the following complex spaces V_n of Abelian integrals $I_\omega = \int_{H=h} \omega$:

(i) $H = y^2 + x^3 - x$, ω-arbitrary complex 1-form of degree $\deg \omega = \max(\deg P, \deg Q) \leq n$, [19], [21];

(ii) $H = x^3 + y^3 + xy$, ω of degree $\leq n$ and symmetric with respect to the transformation $x \to \zeta x$, $y \to \zeta^2 y$, $\zeta^3 = 1$, [19];

(iii) $H = y^2 - x^4 + x^2$, ω of degree $\leq n$, [20];

(iv) $H = y^2 + x^4 - x^2$, ω of degree $\leq n$, [22].

All these Hamiltonians are similar in the following sense. Each of them has 2 critical values $h_1 < h_2$ and the cycle $\gamma = \{H = h\} \cap \mathbf{R}^2$ (in the complex curve $\{H = h\} \subset \mathbf{C}^2$) disappears in one of them, say h_i. Near h_i the integral is holomorphic and the other critical point is a branch point for $I(h)$, $h \in \mathbf{C}$. Their monodromy and Picard-Fuchs equations are also similar.

In all cases presented above, Petrov proves the Chebyshev property of the spaces V_n in three domains:

$D_1 = \{Im h > 0\} \cup (h_1, h_2)$, for the case (i);

$D_2 = \mathbf{C} \setminus (\{h \leq h_1\} \cup \{h_2 \leq h\})$ for the cases, (ii) and (iii) and

$D_3 = \mathbf{C} \setminus (\{h_1\} \cup \{h_2 \leq h\})$ for the cases (i) and (iv).

In the cases (i) and (iii), (iv) with symmetric ω, Petrov considers also the spaces of even and odd integrals (the detailed definitions are in [19]) and shows that they are Chebyshev in the domain D_2.

In the case (iii) he has a stronger result. Let $J(h)$ be the integral of ω along another cycle δ disappearing at $h = h_2$. Then the sum of the number of zeroes of I in D_2 and of the number of zeroes of J in D_2 is less than $\dim V_n$. However, some symmetry assumptions are needed here: either the cycle δ or the form ω is invariant with respect to the central symmetry $S(x,y) = (-x,-y)$. It seems that Petrov missed adding such restriction because without it the results turns to be not true. (The same concerns the symmetry in the case (ii)).

Although the proof of Petrov's theorems are very short they are far from being trivial. They use the Picard-Fuchs equations, the monodromy group and the argument principle for estimating the number of zeroes in the domains D_i.

Let us discuss now some consequences of Petrov's estimates for the Weakened 16-th Hilbert Problem, where the integrals should be considered in the real domain.

1) In the case (i) only for $h_1 \leq h < h_2$ the curve $H = h$ has a compact component and by [**19**] or [**21**] we get the Chebyshev property of V_n in (h_1, h_2).

2) In the case (ii) compact ovals of the curves $H = h$ exist for $h_1 \leq h < h_2$ and by [**20**] we have the Chebyshev property in (h_1, h_2).

3) In the case (iv) the real curves $H = h$ are empty sets for $h < h_1 = -1/4$, consist of two components for $-1/4 \leq h < h_2 = 0$, and form one component for $h > 0$. By [**20**] and [**22**] the Chebyshev property in $(-1/4, 0)$ holds (for symmetric as well as nonsymmetric forms). It is not so in $(-1/4, \infty)$.

Let ω be symmetric, $S^*\omega = \omega$. If $h > 0$ then the real integral I_ω forms the real part of some complex function $J(h)$, which is real (and equal to I_ω) for $-1/4 \leq h \leq 0$ and not real for $h > 0$. The results of Arnold [**1**] and Iliashenko [**14**] suggest that the accuracy of Chebyshev property is 1. In fact, we are far from the Chebyshev property: the dimension of V_n is $2[(n-1)/2]+1$ and the number of zeroes is $\leq 3[(n-1)/2]$ and this bound cannot be lowered. The reason for this relies in the fact that the finite Dulac's series $a_0 + \sum_{i=1}^{k} h^i(a_i + b_i \ln|h|^{-1})$ has at most $3k$ zeroes in $\mathbf{R}$, (see [**24**]). However in the interval $(0, \infty)$ the Chebyshev property with accuracy 1 holds. It can be shown using Petrov's method, (see [**24**]).

If ω is not symmetric then it is natural to consider the integrals: $I^+(h)$ – the integral along the right component of the curve $H = h$ for $-1/4\ h < 0$ and $I(h)$ for $h > 0$.

There is an example [30] showing that the Chebyshev property fails in $(-1/4, \infty)$.

(4) There remained an elliptic polynomial, symmetric with respect to S, which has compact ovals $H = h$. It is $H = y^2 + x^4 + x^2$. If the form ω is symmetric then the transformation $(x, y, h) \to (ix, iy, -h)$, $i^2 = -1$, leads to the situation (iii) with $h < 0 = h_1$. Here we cannot apply the result of [20], but repeating Petrov's proof in [22] one obtains the Chebyshev property in $(0, \infty)$ for our case.

4. Deformations of singularities of vector fields

There are two approaches to the qualitive theory of vector fields. One is based on the study of polynomial systems. It is 16-th Hilbert Problem (second part) in $\mathbf{R}^2$. In the other one investigates vector fields on generic families. It is bifurcation theory. The second problem looks more natural and, in fact, it contains the first one. Namely, the blowing-up of singular point of vector field in $\mathbf{R}^{m+1}$ gives a polynomial vector field in a distinguished divisor $\mathbf{R}^m$. Moreover, versal deformations of singular points are polynomial vector fields. Their investigation is the main part in the proof of their versality. Some examples which lead to the study of Abelian integrals are given below.

(1) Bogdanov-Takens singularity. The linear part of vector field at the singular point is nilpotent. Hence the versal deformation has two parameters: $\dot{x} = y$, $\dot{y} = \mu_1 + \mu_2 y + ax^2 + bxy$, $ab \neq 0$, (see [1]). For μ_2 small with respect to μ_1 we have a small perturbation of the Hamiltonian system with $H = (y^2/2) - \mu_1 x - (ax^3/3)$, (here $x \sim |\mu_1|^{1/2}$, $y \sim |\mu_1|^{3/4}$ and then $xy \sim |\mu_1|^{5/4}$ is smaller than the other nonperturbed terms). Abelian integrals appearing in this situation has been investigated firstly by Bogdanov [2]. Petrov's results are applicable here too.

If we assume that $a = 0$ or that $b = 0$ then we obtain codimension 3 singularities. This volume is devoted to its study in the case $a = 0$ and Parts II and III of my work deal with Abelian integrals appearing in deformations of these singularities. Abelian integrals obtained in the case $b = 0$ are those studied by Petrov, (the case (i)).

(2) One zero and a pair of imaginary eigenvalues. If the singular point in $\mathbf{R}^3$ has such a linear part then the normal form of vector field is such that two variables can be separated, (the direction corresponding to the 0 eigenvalue and the radius in the plane of

rotation). Therefore we obtain a vector field in $\mathbf{R}^2$, symmetric with respect to reflection $(x,y) \to (x,-y)$. The versal deformation is the following: $\dot{x} = \mu_1 + \mu_2 x + ax^2 \pm y^2 + bx^3$, $\dot{y} = -2xy$, $y \geq 0$. For μ_2 small with respect to μ_1 we are in the situation of a small perturbation of an integrable system. The first integral is $H = y^a[x^2 \pm (y^2/(a+2)) + (\mu_1/a)]$. The corresponding problem of zeroes of Abelian integrals $\int y^{a-1}(\mu_2 x + bx^3))dy$ has been solved firstly in [**28**]. Another proofs are given in [**6**],[**8**] and [**13**].

(3) Two pairs of imaginary eigenvalues. If the eigenvalues are non-resonant then the problem in $\mathbf{R}^4$ reduces, (via normal form) to the investigation of vector field in $\mathbf{R}^2$ symmetric with respect to two reflections along the coordinate axes. The versal deformation is the following

$$\dot{x} = x(\mu_1 \pm x^2 - y^2), \quad \dot{y} = y(\mu_2 \mp \frac{\alpha+2}{\beta}x^2 + \frac{\alpha}{\beta+2}y^2 + f(x^2,y^2)), \quad x,y \geq 0, \quad (2)$$

where f is some homogenous polynomial of second degree. If μ_2 is close to $-\alpha\mu_1/\beta$ then the system (2) is a small perturbation of a system with first integral $H = x^\alpha y^\beta[\mu_1 \pm x^2/\beta - (y^2/(\beta+2))]$. Abelian integrals appearing in this problem have been completely investigated in [**29**]. Partial results have also been obtained in [**7**] ($\alpha = \beta$). The authors of [**8**] claimed to have found a simpler proof of the uniqueness of limit cycle in system (2), but their work contains an unavoidable mistake.

(4) Periodic orbits with resonances in $\mathbf{R}^3$. Consider a periodic trajectory of a vector field in $\mathbf{R}^3$ with eigenvalues of the Poincaré map equal to $\exp(\pm 2\pi i p/q)$, $p,q \in \mathbf{Z}$. One can choose the coordinates φ (mod 2π) and $z \in \mathbf{C}$ in a neighbourhood of that periodic orbit such that the natural maps $\{\varphi = \varphi_1\} \to \{\varphi = \varphi_2\}$ are of the form $z \to \exp(ip(\varphi_2 - \varphi_1)/q) \cdot z + \ldots$. The first term of this family of maps defines the Seifert foliation. If we average the z-component of the vector field along the leaves of the Seifert foliation then we get a vector field in $\mathbf{C}$ which is invariant under rotation by an angle $2\pi p/q$. The versal deformation of such vector field is following, (see [**1**]):

$$\begin{aligned} &\dot{x} = y, \quad \dot{y} = \mu_1 + \mu_2 x + ax^2 + bxy, && q = 1, \\ &\dot{x} = y, \quad \dot{y} = \mu_1 x + \mu_2 y + ax^3 + bx^2 y, && q = 2, \\ &\dot{z} = \mu z + Az\,|z|^2 + B\bar{z}^{\,q-1}, && q \geq 3. \end{aligned} \quad (3)$$

The case $q = 1$ was discussed in (1). In the case $q = 2$ we have a small perturbation of the Hamiltonian system with $H = (y^2/2) - (\mu_1 x^2/2) - (ax^4/4)$. We see that this is the situation analyzed in the section 3. For higher q's only the cases $q = 3$ and $q = 4$ are interesting, (if $q \geq 5$ then z^{q-1} is small with respect to $z|z|^2$). Notice that for μ and A pure imaginary the divergence of the vector field (3) vanishes, $\text{div}(\dot{z}) = Re(\partial\dot{z}/\partial z) = 0$. Therefore we have another case of the Weakened 16-th Hilbert Problem. Abelian Integrals in these cases has been investigated by Iliashenko [**14**], for $q = 3$, and by Neishtad [**18**], for $q = 4$.

These are the main applications of Abelian integrals. The higher order singularities are very complicated and the part of their investigation using Abelian integrals is more restricted and difficult. It seems that some generalizations are needed.

The most accesible problem in this field, which probably can be solved, is the Weakened 16-th Hilbert Problem in the class of quadratic systems [**27**]. One result in this direction has been obtained by Zhang, van Gils and Drachman [**31**].

PART II. THE SADDLE AND ELLIPTIC CASES

1. Introduction

R. I. Bogdanov [2] investigated the unfolding

$$\dot{x} = y, \quad \dot{y} = \mu_1 + \mu_2 x + x^2 \pm xy \tag{1}$$

of the codimension two singularity of planar vector field with nilpotent linear part. He proved the topological versality of the family (1). The definition of versality can be found in [1].

The following stage of the local bifurcation theory of differential equations is to investigate the deformations of the codimension three singularities with nilpotent linear part. In fact the other codimension three singularities can be reduced to the study of one dimensional cases.

According to the classification proposed by Dumortier [9], the codimension three singularities with nilpotent linear part are as follows:

$$\begin{aligned} &y\partial/\partial x + (x^2 \pm x^3 y)\partial/\partial y && \text{(cusp case)} \\ &y\partial/\partial x + (\alpha x^3 + xy \pm x^2 y)\partial/\partial y, && \alpha \neq 0, \quad -1/8, \end{aligned} \tag{2}$$

where the case $\alpha > 0$ is called saddle, the cases $-1/8 < \alpha < 0$ and $\alpha < -1/8$ are called respectively elliptic and focus.

The unfolding of the cusp singularity has been fully investigated by Dumortier, Roussarie and Sotomayor [10]. See also the work of Berezovskaya and Khibnik [5].

The investigation of the corresponding elliptic integral has also been done by Yakovlenko [26] and by Petrov [19], who solved the problem of the number of zeroes of the Abelian integrals $\int_{H=h} \omega$ with $H = y^2 - x^3 + x$, (see Part I).

The deformation of the vector field (2) is the following

$$\dot{x} = y, \quad \dot{y} = \mu_1 + \mu_2 x + \alpha x^3 + y(\mu_3 + x \pm x^2). \tag{3}$$

Medved in [17] has considered the family (3) but his bifurcation diagrams are not complete and only partially correct. Bifurcation diagrams in the saddle and focus cases have been proposed by Basikin, Kuznietzov and Khibnik [3], without proofs. The complete investigation of the unfolding (3) has been carried out by Dumortier, Roussarie and Sotomayor [11].

The aim of the present work is to study the family (3) in the region where it is close to a conservative system. The points of codimension two Hopf bifurcation have coordinates $\mu_1 = \mu_3 = 0$, ($\mu_2 < 0$ for the saddle and elliptic cases). One can use the rescaling

$$\tilde{x} = |\mu_2|^{-1/2}x, \quad \tilde{y} = |\mu_2|^{-1}y, \quad \tilde{t} = |\mu_2|^{1/2}t$$

and obtain the system V_β

$$\dot{x} = y, \quad \dot{y} = \sigma x + \alpha x^3 + xy + \beta_0 + \beta_1 y + \beta_2 x^2 y, \quad \sigma = \pm 1 \tag{4}$$

where $\beta_0 = \mu_1|\mu_2|^{-3/2}$, $\beta_1 = \mu_3|\mu_2|^{-1/2}$, $\beta_2 = \pm|\mu_2|^{1/2}$ are small in the domain considered; (we omit the tildas). Observe that for $\beta_i = 0$ the system (4) admits the symmetry $(x,t) \to (-x,-t)$. From this follows that the system V_0 has a first integral H, which we shall find later.

Therefore, system (4) is a small perturbation of a conservative system. Its investigation leads to the study of limit cycles, which bear from the closed curves $H = h$. If the curve $H = h$ is not too close to the separatrix contour or not too big then the problem reduces to the investigation of the following abelian integral, (see [1] for example)

$$J(h) = \Delta H = \int_{H=h} \frac{\partial H}{\partial y}\frac{dx}{\dot{x}}(\beta_0 + \beta_1 y + \beta_2 x^2 y) = \Sigma \beta_i J_i. \tag{5}$$

Every zero of the function J corresponds to the limit cycle. Its stability is determined by the derivative of the function J at this zero.

The main result of this work is the following.

THEOREM 1. *If $\alpha \in (-1/8, \infty)\setminus\{0\}$ then the integral (5) has at most two zeroes counting with multiplicities.*

In the proof of this result we develop a new method, which should be applicable to other problems involving small order perturbation of systems with a complicated

Hamiltonian. In the case of Hamiltonian $y^2 - x^3 + x$ one has a second order Picard-Fuchs equation. Using complex analytic geometry and some estimates Petrov [**19**] has found the best possible estimates of the number of zeroes of the integral $\int_{H=h} \omega$ with ω of arbitrary degree. See also the estimates proposed by Roussarie [**23**]. In the present case ω has three parameters but H is (in general) not algebraic and hence the Picard-Fuchs system turns out to be infinite. We shall find a system of differential equations for certain combinations Q_j of the functions J_i of the form

$$\dot{h} = u(h), \quad \dot{Q}_1 = w(Q, h) + Q_1 R, \quad \dot{Q}_2 = v(Q, h) + Q_2 R \tag{6}$$

with u, w and v quadratic and an unknown function R. It turns out that the function R is not involved in the formula defining the sign of the coefficient A in the expansion $J(h) = A(h - h_*)^3 + \cdots$ at a hypothetical zero h_* of order three of the function J. The sign of A depends on $\dot{Q}_i$ and some estimates about $\dot{Q}_i$ are needed. In the case $\alpha \geq -3/25$ we use the author's previous result [**28**] (see also [**6, 13**]), and for $\alpha \in (-1/8, -3/25)$ a proof of the monotonicity of J_0/J_1 has been given.

If $\alpha < -1/8$ then the problem is more complicated and its investigation is postponed to Part II of this paper.

Part of the results of the present work were obtained duning a visit to the Laboratoire de Topologie at University of Dijon to which the author wants to thank for its hospitality and financial support.

The author also wants to express his gratitude to R. Roussarie, I. Varchenko, and A. Jebrane, J. Sotomayor and C. Rousseau for discussions, which helped him to improve the presentation of is results and avoid some mistakes. Thanks are also due to Mrs. Sarah Oordt from the Department of Mathematics of the University of Arizona in Tucson for the careful typing of a previous version of the present work.

2. Preliminary Transformations

A first integral for system V_0 (4) can be easily found. The substitution $z = \alpha x^2 + \sigma$ and the division of V_0 by x gives the linear system

$$\dot{z} = 2\alpha y, \quad \dot{y} = z + y \tag{7}$$

with the first integral

$$H = |z + \lambda_1 y|^{\lambda_2} \cdot |z + \lambda_2 y|^{-\lambda_1}, \tag{8}$$

where $\lambda_{1,2} = \frac{1}{2}(1 \pm \sqrt{8\alpha + 1})$ are the eigenvalues of the system (7). They are real for $\alpha > -1/8$. From Figure 1 it is seen that only for $\sigma = -1$ part of the curves $H = h$ are closed. This is the case of interest for us.

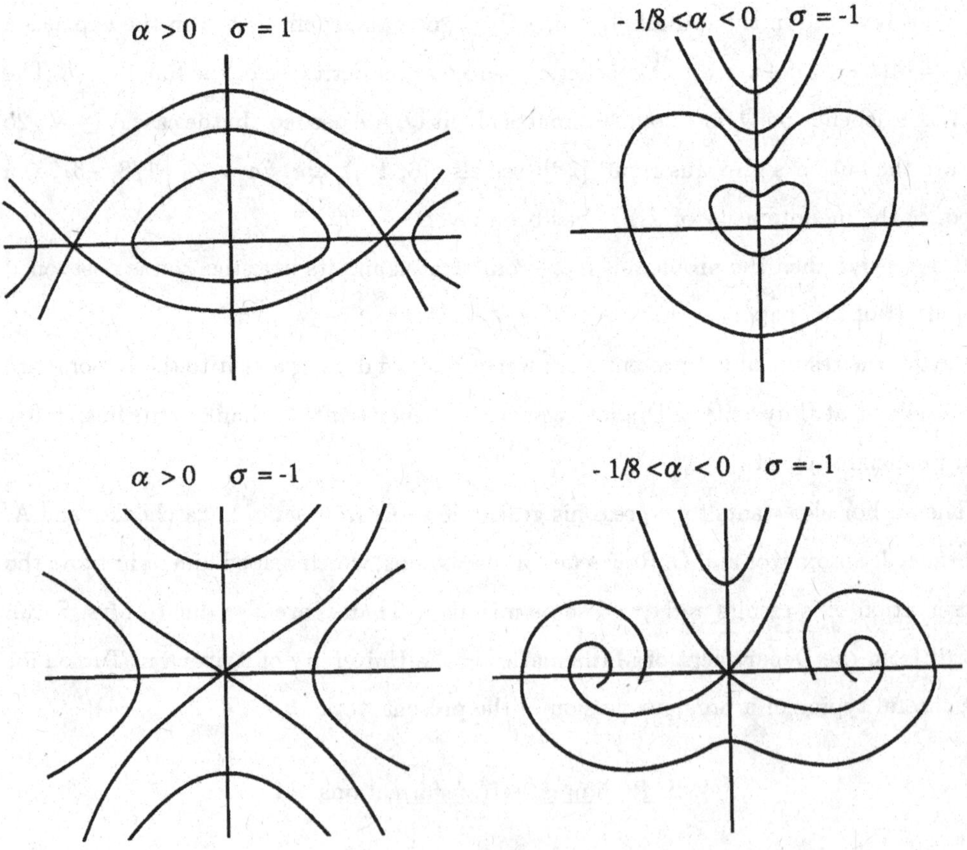

Figure 1

The form (8) of the first integral is not simple enough to our purposes. In order to simplify the problem we shall make some changes of variables.

PROPOSITION 2.1. *The problem of estimation of the number of zeroes of the integral (5) is equivalent to the analogous problem with the function $J(h)$ replaced by*

$$I(c) = \epsilon_0 I_0 + \epsilon_1 I_1 + \epsilon_2 I_2, \tag{9}$$

where

$$I_j = I_j(c) = \int_{F=c} y^{a-2} x^{2j+1} dy, \quad j = 0, 1, 2, \tag{10}$$

$$F = -a(a+1) y^a \left(x^2 + \frac{y}{a+1} + \frac{\sigma}{a}\right), \tag{11}$$

$a \in (-1, 1)\backslash\{0\}$ *is the solution of the equation*

$$a = 2\alpha(a-1)^2 \tag{12}$$

and $c \in (0, 1]$, *for* $a > 0$, *and* $c \geq 1$, *for* $a < 0$.

PROOF: We apply the following change of variables

$$\begin{aligned} x' &= -x/(\sqrt{2}(1-a)) \\ y' &= -(y/(1-a)) - \sigma + (\beta_1 x/(1-a)) - (ax^2/(2(1-a)^2)) + (\beta_2 x^3/(3(1-a))) \\ t' &= t/\sqrt{2}. \end{aligned} \tag{13}$$

The inverse transformation is

$$\begin{aligned} x &= -\sqrt{2}(1-a)x' \\ y &= -((1-a)y') - ((1-a)\sigma) - (\beta_1 \sqrt{2}(1-a)x') \\ &\quad - (a(1-a)x'^2) - (\beta_2 2\sqrt{2}(1-a)^3 x'^3/3). \end{aligned}$$

We obtain the system

$$\begin{aligned} \frac{dx'}{dt'} &= \sigma + ax'^2 + y' + \gamma_1 x' + \gamma_2 x'^3 \\ \frac{dy'}{dt'} &= -2x'y' - \delta_0 - \delta_1 x'^2 - \delta_2 x'^4. \end{aligned} \tag{14}$$

The constants in (14) are as follows

$$\begin{aligned}\gamma_1 &= \sqrt{2}\beta_1\\ \gamma_2 &= 2\sqrt{2}\beta_2(1-a)^2/3\\ \delta_0 &= \sqrt{2}\beta_0/(1-a)\\ \delta_1 &= 2\sqrt{2}\beta_1\\ \delta_2 &= 4\sqrt{2}(1-a)^2\beta_2/3.\end{aligned} \tag{15}$$

If $\beta_i = 0$ then system (14) has F as a first integral and the integral (5) reduces to

$$\Delta F = a(a+1)(\Sigma\gamma_j K_j + \Sigma\delta_j L_j), \tag{16}$$

where

$$\begin{aligned}K_j &= \int_{F=c} y^{a-1}x^{2j-1}dy\\ L_j &= \int_{F=c} y^{a-1}x^{2j}dx.\end{aligned} \tag{17}$$

LEMMA 2.1. *a) For $a \in (-1,1)\backslash\{0\}$ and $\sigma = -1$ all the functions I_j (see (10)) and K_j are positive;*

b) $L_j = \frac{1-a}{2j+1}I_j$;

c) $K_j = -\sigma I_{j-1} - \frac{(2j-1)a+2}{2j+1}I_j$.

PROOF: a) These functions are positive by definition because we choose the orientation of the curve $F = c$ defined by the direction opposite to the vector field (14).

b) It follows from the identity

$$y^{a-1}x^{2j}dx = d\left(\frac{y^{a-1}x^{2j+1}}{2j+1}\right) - \frac{a-1}{2j+1}y^{a-2}x^{2j+1}dy.$$

c) On the curve $F = c$, one has

$$2xydx + (ax^2 + y + \sigma)dy = 0.$$

From this we obtain

$$\begin{aligned} K_j + \sigma I_{j-1} &= \int y^{a-2}x^{2j-1}(y+\sigma)dy \\ &= -\int y^{a-2}x^{2j-1}(ax^2dy + 2xydx) \\ &= -aI_j - \frac{2}{2j+1}\int y^{a-1}d(x^{2j+1}) \\ &= -\frac{(2j+1)a + 2(1-a)}{2j+1}I_j. \blacksquare \end{aligned}$$

Form (15), (16) and Lemma 2.1 we obtain that the problem of estimation of the number of zeroes of the function J reduces to the analogous problem with I, (see (9)), where

$$\begin{aligned} \epsilon_0 &= \sqrt{2}(\beta_0 + \beta_1) \\ \epsilon_1 &= \sqrt{2}a\beta_1 + (2\sqrt{2}(1-a)\beta_2/3) \\ \epsilon_2 &= -2\sqrt{2}a(1-a)^2\beta_2/3. \end{aligned}$$

The latter transformation is nonsingular for $a \in (-1,1)\backslash\{0\}$. From this Proposition 2.1 follows. ∎

Remark 2.1. The Hamiltonian (11) has appeared in the study of codimension 2 singularity of planar vector fields symmetric with respect to y-axis, (see [**28**]). From the proof of Proposition 2.1 it follows that there it is a symmetry $a \to 1/a$. This symmetry was not discovered in [**11**]; it relies upon the exchange of the parabolas $y = 0$ and $ax^2 + y - 1 = 0$.

3. Differential Equations

In this section we shall obtain differential equations satisfied by I_j and by their ratios. These equations contain an unknown function, but (as we shall see) this new function does not influence negatively certain conditions appropriate for solving the problem of zeroes of the integral (9). This will be done in Section 6.

Because of (10) and (11)

$$I_j = (2j+1)\iint_{F\leq c} y^{a-2}x^{2j}dxdy = -\frac{2j+1}{2a(a+1)}\iint_{F\leq c} x^{2j-1}y^{-2}dydF$$

one has

$$\frac{dI_j}{dc} = I_j' = -\frac{2j+1}{2a(a+1)} \int_{F=c} x^{2j-1} y^{-2} dy.$$

Therefore, using integration by parts one can easily find that the following differential equations are satisfied

$$2(a+1)cI_0' = (2a-1)I_0 + \frac{\sigma}{a} I_{-1}$$
$$2(a+1)cI_1' = (2a+1)I_1 + \frac{3\sigma}{a} I_0$$
$$2(a+1)cI_2' = (2a+3)I_2 + \frac{5\sigma}{a} I_1.$$

It is useful to introduce the functions

$$Q_1 = \frac{I_1}{I_0}, \quad Q_2 = \frac{I_2}{I_0}, \quad R = \sigma \frac{c-1}{a} I_{-1}/I_0.$$

By (18) their graph forms a trajectory of the following vector field $X(P)$ in $\mathbf{R}^4 = \{P = (c, Q_1, Q_2, R)\}$

$$\dot{c} = 2(a+1)c(1-c)$$
$$\dot{Q}_1 = 3\frac{1-c}{a}\sigma + 2(1-c)Q_1 + Q_1 R$$
$$\dot{Q}_2 = 5\frac{1-c}{a}\sigma Q_1 + 4(1-c)Q_2 + Q_2 R$$
$$\dot{R} = G(c) \tag{19}$$

where $G(c) = 2(a+1)c(1-c) \cdot dR/dc$ is a given function of c.

Denote

$$\Omega = \{(Q_1, Q_2)(c) : c \in (0,1](a > 0) \quad \text{or} \quad c \geq 1(a < 0)\}. \tag{20}$$

The importance of introducing the variables Q_i is seen from the following lemma whose proof is straightforward.

LEMMA 3.1. *The assertion of Theorem 1 is equivalent to the following three assertions:*

a) Ω is nonsingular ($\Sigma|\dot{Q}_i| > 0$);

b) Ω has no self-intersections;

c) Ω has no inflection points ($\dot{Q} \neq$ constant $\ddot{Q}$).

In other words Ω is an embedded, convex and nonsingular curve in $\mathbf{R}^2$.

4. Asymptotic Behaviour of Q_i

In this section we describe the behaviour of the functions Q_i near the ends of their domains of definition.

Starting at this point we assume that $\sigma = -1$ (see Figure 1).

LEMMA 4.1. *(Behaviour near $c = 1$). a) $R(c) \to const$ as $c \to 1$;*
b) $Q_1(c) = \frac{3(1-c)}{4a(a+1)} + o((1-c))$ as $c \to 1$;
c) $Q_2 = \frac{10}{9}Q_1^2 + o((1-c)^2)$ as $c \to 1$, (Ω is convex near the point $(0,0)$).

PROOF: Since

$$x^2 = \frac{1-c}{a(a+1)} - \frac{1}{2}(y-1)^2 + o(|1-c| + (y-1)^2)$$

on the curve $F = c$ near $(y, c) = (1, 1)$ one has

$$\int y^{a-2}x^{2i+1}dy = \sqrt{2}[\frac{1-c}{a(a+1)}]^{i+1}k_i(1+o(1)), \quad i = -1, 0, 1, 2,$$

where

$$k_i = \int_0^{2\pi} (\sin\alpha)^{2i+2}d\alpha = \begin{cases} \pi, & i = 0 \\ 3\pi/4, & i = 1 \\ 5\pi/8 & i = 2. \end{cases}$$

From this the proof follows easily. ∎

Remark. Notice that the functions Q_i and R are not analytic near $c = 1$.

LEMMA 4.2. *(Behaviour near $c = 0$). Let $a \in (0, 1)$. Then:*
a) $Q_1 \to \frac{3}{a+2}$ as $c \to 0$;
b) As $c \to 0$,

$$Q_2 = \frac{15}{(a+2)(3a+2)} + \frac{10(a+1)^2}{a(2a+3)(3a+2)}(Q_1 - \frac{3}{a+2}) + o(Q_1 - \frac{3}{a+2}) \tag{21}$$

PROOF: Denote, (see (17))

$$S_1 = K_1/I_0, \quad S_2 = K_2/I_0. \tag{22}$$

By Lemma 2.1 we have

$$S_1 = 1 - \frac{a+2}{3} Q_1, \quad S_2 = Q_1 - \frac{3a+2}{5} Q_2. \tag{23}$$

Therefore

$$Q_1 = \frac{3}{a+2}(1 - K_1 I_0^{-1})$$

$$Q_2 = \frac{15}{(a+2)(3a+2)}(1 - (\frac{a+2}{3}K_2 + K_1) I_0^{-1}),$$

where

$$K_1 \to 2\int_0^{(a+1)/a} y^{a-1}(\frac{1}{a} - \frac{y}{a+1})^{1/2} dy = 2a^{-1/2} \cdot (\frac{a+1}{a})^a \cdot B(a, \frac{3}{2}) = D$$

($B(\alpha, \beta)$ is the beta function, see [4]), and

$$K_2 \to 2\int_0^{(a+1)/a} y^{a-1}(\frac{1}{a} - \frac{y}{a+1})^{3/2} dy = \frac{3}{a(2a+3)} D \quad \text{as} \quad c \to 0.$$

On the other hand

$$I_0 = 2a^{-1/2} \int y^{a-2}(1 - \frac{ay}{a+1} - \frac{c}{a+1} y^{-a})^{1/2} dy$$

$$= 2a^{-1/2} \int_{[c/(a+1)]^{1/a}}^{\infty} y^{a-2}(1 - \frac{c}{a+1} y^{-a})^{1/2} dy + 0(1).$$

Using the substitution $z = \frac{c}{a+1} y^{-a}$ one finds $I_0 \to \infty$ as $c \to 0$. From this the proof follows easily. ∎

LEMMA 4.3. *(Behaviour near $c = \infty$). Let $a \in (-1, 0)$. Then:*

a) $Q_1 \to \frac{3}{a+2}$ *as* $c \to \infty$;

b) If $a \in (-2/3, 0)$ *then*

$$Q_2 = \frac{15}{(a+2)(3a+2)} - E(\frac{3}{a+2} - Q_1)^{\gamma} \cdot (1 + o(1)) \quad \text{as}$$

$$c \to \infty, \quad \text{where} \quad E > 0 \quad \text{and} \quad \gamma = (3a+2)/(2a+2) > 0;$$

c) If $a = -2/3$ *then*

$$Q_2 = -F \cdot \ln(\frac{3}{a+2} - Q_1) \cdot (1 + o(1)) \quad \text{as} \quad c \to \infty, \quad \text{where} \quad F > 0;$$

d) If $a \in (-1, -2/3)$ then

$$Q_2 = G \cdot (\frac{3}{a+2} - Q_1)^{\gamma} \cdot (1 + o(1)) \text{ as } c \to \infty, \text{ where}$$

$$G > 0 \text{ and } \gamma = (3a+2)/(2a+2) < 0.$$

Proof: Recall that here $a < 0$ and $c \to \infty$. a) As in the previous proof we use the formulas (22) and (23). We have

$$I_0 = \frac{2}{\sqrt{|a|}} \int y^{a-2} (\frac{c}{a+1} y^{-a} - 1 + \frac{ay}{a+1})^{1/2} dy, \tag{24}$$

where the limits of the integral are $(\frac{c}{a+1})^{1/a} \cdot (1+o(1)) \to 0$ and $|\frac{c}{a}|^{1/(a+1)} \cdot (1+o(1)) \to \infty$. Using the substitution $\frac{c}{a+1} y^{-a} = z^{-1}$ one finds

$$I_0 = 2 \cdot |a|^{-3/2} \cdot (\frac{c}{a+1})^{1-(1/a)} \int z^{(1/2)-(1/a)-1}$$

$$\cdot [1 - z - \frac{|a|}{a+1} (\frac{c}{a+1})^{1/a} \cdot z^{(a+1)/a}]^{1/2} dz$$

$$= \text{const} \cdot c^{1-(1/a)} \cdot (1 + o(1)).$$

The corresponding limits of the integral are $\frac{a+1}{c} |\frac{c}{a}|^{a/(a+1)} \cdot (1 + o(1)) \to 0$ and $1 + o(1)$.

The same substitution works with K_1

$$K_1 = 2|a|^{-3/2} \frac{c}{a+1} \int z^{-1/2} [1 - z - 0(c^{1/a}) \cdot z^{(a+1)/a}]^{1/2} dz$$

$$= \text{const} \cdot c \cdot (1 + 0(1)).$$

From this and from (22) and (23) part a) of Lemma 4.2 follows.

b) Let $a \in (-2/3, 0)$. By a) and by (22) and (23) it suffices to show that

$$K_2 \sim \text{const } c^{\delta}, \quad \delta = \frac{2a+3}{2a+2} > 1.$$

Then $K_2 I_0^{-1} \sim \text{const } (K_1 I_0^{-1})^{\gamma}$, where $K_1 I_0^{-1} \sim \text{const} \cdot c^{1/a}$ and $\gamma = \frac{3a+2}{2a+2}$. One has

$$K_2 = 2(a+1)^{-3/2} \int y^{a+(1/2)} [|\frac{c}{a}| y^{-a-1} - 1 - \frac{a+1}{|a|} y^{-1}]^{3/2} dy,$$

the limits are as in (24). We use the substitution $(c/|a|)y^{-a-1} = z^{-1}$. Then

$$K_2 = 2(a+1)^{-5/2}|\frac{c}{a}|^{\delta} \int z^{(-a/(2a+2))-1}(1 - z - \frac{a+1}{|a|}$$

$$\cdot |\frac{a}{c}|^{1/(a+1)} z^{a/(a+1)})^{3/2} dz, \tag{25}$$

the limits are $|\frac{a}{c}|(\frac{c}{a+1})^{(a+1)/a} \cdot (1 + o(1))$ and $1 + o(1)$. From this we get part b) of Lemma 4.3.

c) Let $a = -2/3$. Then

$$I_2 = 2(\frac{2}{3})^{-5/2} \cdot \int_{\sim(3c)^{-3/2}}^{\sim(3c/2)^3} (3c)^{5/2} \cdot y^{-1} \cdot (1 - \frac{y^{-2/3} + 2y^{1/3}}{3c})^{5/2} dy$$

$$= 2(\frac{9c}{2})^{5/2} \cdot [\int_{(3c)^{-3/2}\sqrt{\ln c}}^{(3c/2)^3/\sqrt{\ln c}} y^{-1}(1 - \frac{y^{-3/2} + 2y^{1/3}}{3c})^{5/2} dy + 0(\sqrt{\ln c})]$$

$$= 9(\frac{9c}{2})^{5/2} \cdot \ln(c) \cdot (1 + o(1)).$$

By part a)

$$I_0 \sim \text{const } c^{5/2}$$

which ends this part of Lemma 4.3.

d) Let $a \in (-1, -2/3)$. Then

$$I_2 = 2 \cdot (|a|(a+1))^{-5/2} \cdot c^{5/2} \int y^{-(3a/2)-2} \cdot (1 - \frac{|a|}{c} y^{a+1} - \frac{a+1}{c} y^a)^{5/2} dy,$$

the limits are as in (24). After the substitution $(|a|/c)y^{a+1} = z$ we get

$$I_2 = \text{const } c^{(5/2)-\gamma} \cdot \int z^{-\gamma-1} \cdot (1 - z - \frac{a+1}{c}|\frac{c}{a}|^{a/(a+1)} \cdot z^{a/(a+1)})^{5/2} dz,$$

$$\gamma = (3a+2)/(2a+2) < 0$$

(the limits are as in (25)), and one can easily see that

$$I_2 \sim \text{const } c^{(5/2)-\gamma},$$

$$I_2/I_0 \sim \text{const } (c^{1/a})^{\gamma},$$

where $\frac{3}{a+2} - Q_1 \sim K_1 I_0^{-1} \sim \text{const } c^{1/a}$. ∎

5. Regularity and Embeddness of Ω

Let us consider the transformation $(Q_1, Q_2) \to (S_1, S_2)$ given by formula (23). In other words we have performed an affine change of the variables Q_1 and Q_2. In [28] (see also [**6**, **8**, **13**]) the following result has been proven.

THEOREM 2. *The ratio $S_2/S_1 = K_2/K_1$ is strictly monotone.*

In terms of the variables Q_1 and Q_2 the assertion of Theorem 2 means that the coefficient of the slope of the line passing through the points $(\frac{3}{a+2}, \frac{15}{(a+2)(3a+2)})$ and $Q(c)$ is monotone; for $a = -2/3$ we obtain the monotonicity of Q_1. From this the regularity and embeddness of the curve Ω follows.

6. Convexity of Ω

By Lemma 4.1.c) the curve Ω is convex near its endpoint (0,0). Let $\overline{Q} = (\overline{Q}_1, \overline{Q}_2)$ be the inflection point of Ω nearest to $(0,0)$, along Ω, which we hypothetically assume that exists.

LEMMA 6.1. a) *$\Omega = \{(Q_1, W(Q_1))\}$ near $\overline{Q}$;*
b) $\dot{Q}_1 < 0$ near $\overline{Q}$;
c) $0 < W'(\overline{Q}_1) < W'(\frac{3}{a+2})$, $W''(\overline{Q}_1) = 0$, $W'''(\overline{Q}_1) \leq 0$;
d) if $W'''(\overline{Q}_1) = 0$
then there is a trajectory of the system (19) close to our trajectory, whose projection, $\tilde{\Omega}$, onto the Q-plane has an inflection point $\tilde{Q}_1$ satisfying $\tilde{W}'''(\tilde{Q}_1) < 0$.

PROOF: If $a = -2/3$ then Theorem 2 means the monotonicity of Q_1 and the properties a) and c) are obvious because $W'(3/(a+2)) = \infty$ for $a \in (-1, 0)$ (see Lemma 4.3c).

Let $a \in (-2/3, 1)\backslash\{0\}$. Since $K_i > 0$ and $I_j > 0$ (see Lemma 2.1) the functions Q_i and S_i are positive. Hence Ω is contained in the triangle with vertices: (0,0), $(3/(a+2), 0)$ and $(\frac{3}{a+2}, \frac{15}{(a+2)(3a+2)})$. From Figure 2 it is seen that $\dot{Q}_2 < 0$ at the first point of verticality of Ω between (0,0) and $\overline{Q}$. Hence there should exist an extremal point of S_2/S_1, (see remarks after Theorem 2). Therefore $\dot{Q}_1 < 0$ and it remains to prove that $W'(Q_1) < W'(\frac{3}{a+2})$. By the monotonicity of S_2/S_1 the curve Ω lies between the line connecting its enpoints and the line tangent to Ω in its right enpoint. If $W'(\overline{Q}_1) \geq W'(3/(a+2))$ then the right endpoint $(\frac{3}{a+2}, W(\frac{3}{a+2}))$ lies below the line tangent to Ω at $\overline{Q}$. This means that the sign at the derivative $\frac{d}{dc}(S_2/S_1)$ at $\overline{Q}$ is opposite to the sign of this derivative at $(0,0)$.

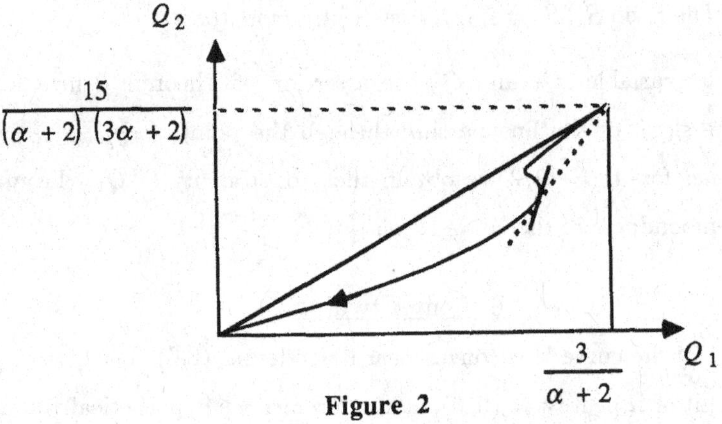

Figure 2

The point d) is clear if we observe that by (19) the condition $W''' = 0$ defines a hypersurface in $\mathbf{R}^4$, which does not coincide with the hypersurface $W'' = 0$.

If $a \in (-1, -2/3)$ then the desired result is a consequence of the following theorem, the proof of which we postpone to the next section.

THEOREM 3. *$dQ_1/dc > 0$ for $a \in (-1, 0)$.*

Remark 6.1. From the proof of Theorem 3 presented here it follows also that

$$dQ_1/dc \neq 0, \quad dQ_2/dc \neq 0 \quad \text{and} \quad d(Q_2/Q_1)/dc \neq 0$$

for all $a \in (-1,1)\backslash\{0\}$.

We continue the proof of the convexity of Ω. With any point Q of Ω we associate the line $L : Q_2 = \lambda Q_1 + \omega$ tangent to Ω at Q. If the point Q is the inflection point of Ω then L has a tangency of order 2.

Remark. The curve $\hat{\Omega} = \{L : L \text{ tangent to } \Omega\}$ is the dual curve and each inflection point of the curve Ω corresponds to the cusp point of $\hat{\Omega}$.

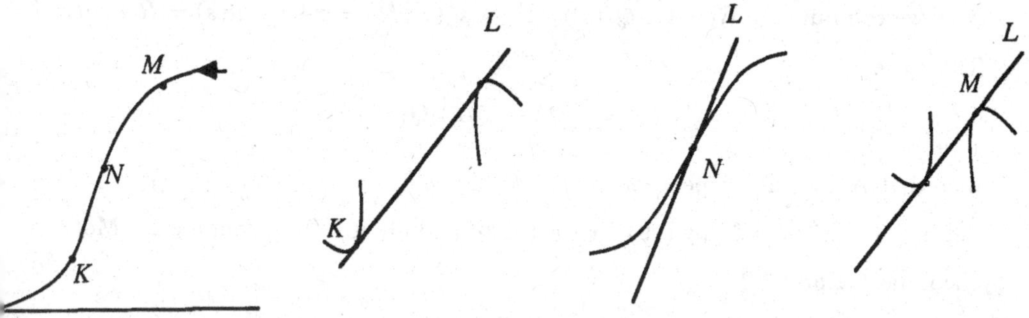

Figure 3

The line L defines the three dimensional hyperplane in $\mathbf{R}^4$

$$M = M_{\lambda,\omega} = \{(c, Q_1, Q_2, R) : Q_2 = \lambda Q_1 + \omega\}.$$

Through any point $P \in \mathbf{R}^4$ passes a line $Z = \{P + X(P)s : s \in \mathbf{R}\}$ oriented by the vector field X (19).

We look at the points on the line $Z \subset M$, at which the vector field X is tangent to M. More precisely, we consider the function

$$\zeta(s) = (\dot{Q}_2 - \lambda\dot{Q}_1)|_Z.$$

The function

$$(\dot{Q}_2 - \lambda\dot{Q}_1)|_M = (4w + 3\lambda/a)(1 - c) + wR + (2\lambda - 5/a)Q_1(1 - c)$$

is quadratic on M, parametrized by c, Q_1 and R. Therefore ζ is also quadratic

$$\zeta(s) = As^2 + Bs$$

and hence has two zeroes (with multiplicities). Each zero corresponds to the different kind of contact of the integral curves of X with M at these points. At the inflection point $B = 0$ and the character of tangency is determined by the sign of the coefficient A.

If we are in the situation described by Lemma 6.1 then $A > 0$ and the situation is presented in Figure 3, where the projections of trajectories tangent to M are illustrated.

Now we compute A. We have $Q_1(s) = Q_1 + s\dot{Q}_1$, $c(s) = c + s\dot{c}$, $R(s) = R + s\dot{R}$. Therefore

$$A = -(2\lambda - (5/a))\dot{Q}_1\dot{c}.$$

Notice that A does not depend on R and $\dot{R}$.

Let $a > 0$. Then $\dot{c} > 0$ (by (19) for $c \in (0,1)$) and $\dot{Q}_1 < 0$ (by Lemma 6.1). Moreover by Lemmas 6.1 and 4.2

$$\lambda = W'(\overline{Q}_1) < W'(3/(a+2)) = \frac{10(a+1)^2}{a(2a+3)(3a+2)}.$$

One can easily check that $W'(3/(a+2)) < 5/(2a)$. Therefore $A < 0$ and we have a contradiction.

Let $a < 0$. Then $\dot{c} < 0$, $\dot{Q}_1 < 0$, $2\lambda - (5/a) > 0$ and hence $A < 0$. This completes the proof of the convexity of Ω.

7. Monotonocity of Q_1

In this section we shall prove Theorem 3. But instead of estimating the derivative of Q_1 we shall show that

$$S_1' = dS_1/dc < 0.$$

By (23) these two facts are equivalent.

We have

$$I_0 = 2\int_{\overline{y}}^{\overline{\overline{y}}} y^{a-2}x\,dy, \quad I_0' = -\frac{2}{a(a+1)}\int_{\overline{y}}^{\overline{\overline{y}}} y^{-2}x^{-1}dy,$$

$$K_1 = 2\int_{\overline{y}}^{\overline{\overline{y}}} y^{a-1}x\,dy, \quad K_1' = -\frac{2}{a(a+1)}\int_{\overline{y}}^{\overline{\overline{y}}} y^{-1}x^{-1}dy,$$

where $\overline{y} < \overline{\overline{y}}$ are the points of the intersection of the curve $F = c$ with the line $x = 0$ and $x = [(1/a) - (y/(a+1)) - (cy^{-a}/a(a+1))]^{1/2}$. Therefore we have to show that

$$\iint \Phi(y_1, y_2)dy_1dy_2 < 0,$$

where

$$\Phi(y_1, y_2) = \frac{y_2^{a-2}x_2}{y_1x_1} - \frac{y_1^{a-1}x_1}{y_2^2x_2}.$$

Obviously it is sufficient to show that

$$\iint \Psi(y_1, y_2) dy_1 dy_2 < 0, \tag{26}$$

where

$$\Psi(y_1, y_2) = \Phi(y_1, y_2) + \Phi(y_2, y_1).$$

We have

$$\Psi(y_1, y_2) = y_1^{(a/2)-2} \cdot y_2^{(a/2)-2} \cdot (y_2 - y_1) \frac{\phi(y_1) - \phi(y_2)}{\sqrt{\phi(y_1)\phi(y_2)}}, \tag{27}$$

where $\phi(y) = \frac{y^a}{a} - \frac{y^{a+1}}{a+1} - \frac{c}{a(a+1)}$. The graph of the function ϕ is represented in Figure 4. We shall reduce the integral along $[\overline{y}, \overline{\overline{y}}]^2$ to an integral along $[\overline{y}, 1]^2$. Namely, with each $y_i < 1$ we associate $y_i' > 1$, $i = 1, 2$ satisfying $\varphi(y_i) = \varphi(y_i') = v_i$, (see fig. 4). Then we change the integration along y_i' to the integration along y_i. One can easily check that the desired inequality (26) follows from the next result.

Remark. The author ows to R. Roussarie the remark about the necessity of the derivatives dy_i'/dy_i in (28), omitted in an earlier version of this work.

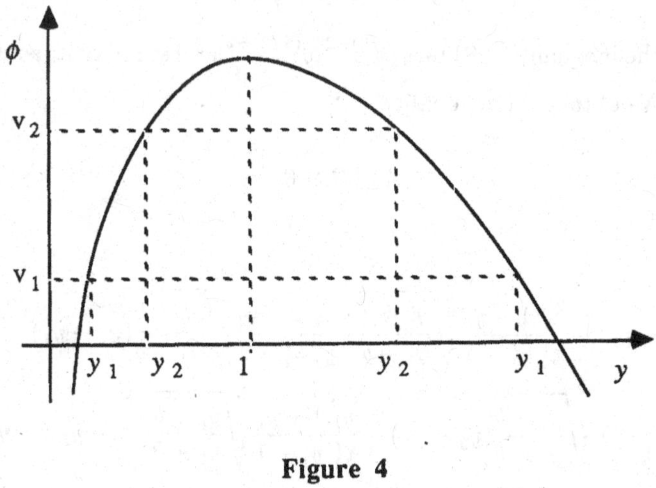

Figure 4

LEMMA 7.1. *We have the following inequality*

$$y_1^{(a/2)-2}\cdot y_2^{(a/2)-2}\cdot(y_2-y_1)+y_1'^{(a/2)-2}\cdot y_2'^{(a/2)-2}\cdot(y_2'-y_1')\cdot\frac{dy_1'}{dy_1}\cdot\frac{dy_2'}{dy_2}$$
$$-y_1^{(a/2)-2}\cdot y_2'^{(a/2)-2}\cdot(y_2'-y_1)\frac{dy_2'}{dy_2}$$
$$-y_1'^{(a/2)-2}\cdot y_2^{(a/2)-2}\cdot(y_2-y_1')\frac{dy_1'}{dy_1}>0. \tag{28}$$

PROOF: Firstly we give some relations between y_i's and y_i''s. The equation $\phi(y)=\phi(y')$ means that

$$\left(\frac{y}{y'}\right)^a=\frac{y'+b}{y+b} \tag{29}$$

or that

$$by^{-1/(b+1)}+y^{b/(b+1)}=by'^{-1/(b+1)}+y'^{b/(b+1)} \tag{30}$$

where $0<y<1<y'$ and

$$b=-(a+1)/a>0$$

for $a\in(-1,0)$. $y'=y'(y)$ is the function of y. Its derivative is

$$\frac{dy'}{dy}=\left(\frac{y}{y'}\right)^{a-1}\frac{y-1}{y'-1}=\frac{y'}{y}\cdot\frac{y'+b}{y+b}\cdot\frac{y-1}{y'-1}<0. \tag{31}$$

Let us divide the inequality (28) by $y_1'^{(a/2)-2}\cdot y_2^{(a/2)-2}$ and simplify it using the formulas (29) and (31). We obtain the inequality

$$A+B>0,$$

where

$$A=(y_2-y_1)-\frac{y_1}{y_1'}\frac{1-y_1}{y_1'-1}\sqrt{\frac{y_1'+b}{y_1+b}}\cdot\frac{y_2}{y_2'}\frac{1-y_2}{y_2'-1}\sqrt{\frac{y_2'+b}{y_2+b}}(y_1'-y_2') \tag{32}$$

$$B=\frac{y_2}{y_2'}\frac{1-y_2}{y_2'-1}\sqrt{\frac{y_2'+b}{y_2+b}}(y_2'-y_1)-\frac{y_1}{y_1'}\frac{1-y_1}{y_1'-1}\sqrt{\frac{y_1'+b}{y_1+b}}(y_1'-y_2),\quad y_1<y_2.$$

We shall show the inequalities

$$A>0,\qquad B>0.$$

The formulas (32) are not very useful to us because we do not know how to estimate the ratios $(y_1' - y_2')/(y_2 - y_1)$ and $(y_1' - y_2)/(y_2' - y_1)$ from above, (see Figure 4). Fortunately, there is a symmetry, established in 7.2 which allows us to transform our problem to the simpler one. In the transformed inequalities we will have to estimate the corresponding ratios from below. It will be done in Lemma 7.3.

LEMMA 7.2. *If $y < y'$ are the solutions of the equation (30) with the parameter b then $\tilde{y} = 1/y'$ and $\tilde{y}' = 1/y$ are the solutions of this equation with the parameter $\tilde{b} = 1/b$.*

After applying this symmetry and performing some transformations we get the following inequalities to prove, (we omit the tildas):

$$(y_1' - y_2') \geq \psi(y_1)\psi(y_2)(y_2 - y_1),$$
$$\psi(y_2)(y_1' - y_2) \geq \psi(y_1)(y_2' - y_1), \quad y_1 < y_2,$$

where

$$\psi(y) = \frac{y' - 1}{\sqrt{y'(y' + b}} / \frac{1 - y}{\sqrt{y(y + b}}.$$

The result follows from the following two lemmas.

LEMMA 7.3. *a) $y_1' - y_2' \geq y_2 - y_1$;*
b) $y_1' - y_2 \geq y_2' - y_1$.

LEMMA 7.4. *$\psi(y)$ is an increasing function of y and $0 \leq \psi(y) \leq 1$.*

PROOF OF LEMMA 7.3: We have $y_1' - y_2 = (y_1' - y_2') + (y_2' - y_2)$ and $y_2' - y_1 = (y_2 - y_1) + (y_2' - y_2)$, (see Figure 4). Therefore it is enough to show the point a) or the inequality

$$dy'/dy \leq -1.$$

If $y \to 1$ then $y' \to 1$,

$$\frac{y' - 1}{1 - y} = 1 + \frac{2}{3}\frac{b + 2}{b + 1}(1 - y) + \cdots, \tag{33}$$

and from (31) we get $(dy'/dy)(1) = -1$. (The property (33) follows from the expansion of the equation (30) near $y = 1$, $y' = 1$.) Now, differentiating the identity (31) we obtain

$$\frac{d^2y'}{dy^2} = \frac{y'(y' + b)}{y' - 1}\left(\frac{1 - y}{y(y + b)}\right)^2(1 + b)[(1 - y)^{-2} - (y' - 1)^{-2}].$$

This expression is nonnegative since $k(y) = (y' - 1) - (1 - y) \geq 0 : k(0) = \infty$, $k(y) = \frac{2}{3}\frac{b+2}{b+1}(1 - y)^2 + \cdots$ for $y \to 1$ (by (33)) and $dk/dy\Big|_{k=0} = 1 - (y/y')^{a-1} \leq 0$ for $a < 0$. Now the inequality a) of Lemma 7.3 is obvious. ∎

PROOF OF LEMMA 7.4: We have

$$\frac{d(\psi^2)}{dy} = \frac{(y' - 1)^2}{y'(y' + b)(1 - y)} 2(b + 1)(u + u')x(y), \quad x(y) = u - u' - \frac{b + 2}{2(b + 1)},$$

where $u = 1/(1 - y) > u' = 1/(y' - 1) \geq 0$, $u \geq 1$. If $y = 0$ then $u = 1$, $u' = 0$ and $x(0) > 0$. For $y \to 1$ $u = u'(1 + \frac{2}{3}\frac{b+2}{b+1}\frac{1}{u} + \cdots)$ and hence $x(1) > 0$. Next we calculate the derivative of x

$$\frac{dx}{dy} = \frac{1 - y}{y(y + b)}\{(u - u')[1 + (b + 1)(u^2 + uu' + u'^2)] - (b + 2)(u^2 + u'^2)\}.$$

Therefore the sign of $(dx/dy)\Big|_{x=0}$ is equal to sign of

$$\lambda(u) = -4(b + 1)^2u^2 + 2(b + 1)(b + 2)u - b^2, \quad u \geq 1.$$

One can easily check that $\lambda(u) < 0$ for $u \geq 1$ (and $b > 0$). This completes the proof of Lemma 7.4 and of Theorem 3. ∎

PART III. THE FOCUS CASE

1. Introduction

This is a continuation of the work initiated in Part II, concerning the system

$$\dot{x} = y, \quad \dot{y} = \mu_1 + \mu_2 x + \alpha x^3 + \mu_3 y + xy \pm x^2 y. \tag{1}$$

which was analyzed for $\alpha \in (-1/8, \infty)\backslash\{0\}$ near the ray $\mu_1 = \mu_3 = 0$, $\mu_2 < 0$. Now we will deal with $\alpha < -1/8$ (the focus case) near the two rays $\mu_1 = \mu_3 = 0$, $\mu_2 < 0$ and $\mu_1 = \mu_3 = 0$, $\mu_2 > 0$.

The normalization $x \to x/\sqrt{|\mu_2|}$, $y \to y/|\mu_2|$, $t \to \sqrt{|\mu_2|}t$ leads to the system

$$\dot{x} = y, \quad \dot{y} = \pm x + \alpha x^3 + xy + \beta_0 + \beta_1 y + \beta_2 x^2 y \tag{2}$$

For certain future purpose, together with system (2), we consider also the system

$$\dot{x} = y + \beta_{-1} x^{-1}, \quad \dot{y} = \sigma x + \alpha x^3 + xy + \beta_0 + \beta_1 y + \beta_2 x^2 y, \quad \sigma = \pm 1. \tag{3}$$

If $\beta_j = 0$ system (3) has the first integral

$$H = [\frac{-\sigma - \alpha x^2 - \lambda_1 y}{-\sigma - \alpha x^2 - \overline{\lambda}_1 y}]^{i/2b} \cdot [(\sigma + \alpha x^2 + \frac{y}{2})^2 + (\frac{by}{2})^2]^{-1/2}. \tag{4}$$

Here $\lambda_{1,2} = \frac{1}{2}(1 \mp ib)$, $b = \sqrt{|8\alpha + 1|}$, $i = \sqrt{-1}$. In the perturbed system (3) most of the closed curves $H = h$ disappear and only from few of them limit cycles are born. Its investigation is directly connected with the study of the zeroes of the following Abelian integral

$$\begin{aligned} J = \sum \beta_i J_i = \beta_{-1} \int_{H=h} \frac{\partial H}{\partial x} \cdot \frac{1}{x} \cdot \frac{dy}{\dot{y}} \\ + \int_{H=h} \frac{\partial H}{\partial y} \frac{dx}{\dot{x}} (\beta_0 + \beta_1 y + \beta_2 x^2 y), \end{aligned} \tag{5}$$

where

$$\frac{\partial H}{\partial x} = -2\alpha H R^{-2} \dot{y}, \quad \frac{\partial H}{\partial y} = 2\alpha H R^{-2} \dot{x}, \tag{6}$$

and

$$R^2 = (\sigma + \alpha x^2 + \frac{1}{2}y)^2 + (\frac{by}{2})^2, \quad \dot{x} = y, \quad \dot{y} = x(\sigma + \alpha x^2 + y).$$

The main result of this work is the following.

THEOREM 1. *The function*

$$h \to \beta_0 J_0 + \beta_1 J_1 + \beta_2 J_2$$

has at most two zeroes counting their multiplicities.

2. Some Formulas and Properties of Integrals

The form (4) of the first integral is not very useful for calculations with abelian integrals. Moreover it is not uniquely defined. In order to simplify it we introduce some notations.

Let

$$u = 1 + \sigma(\alpha x^2 + \frac{y}{2}), \quad v = \sigma by/2, \quad R = \sqrt{u^2 + v^2} \quad \text{and} \quad \phi \tag{7}$$

be such that $tg\phi = v/u$; ϕ is uniquely defined in the domain $\mathbf{R}^2 \backslash \{(u, v) : u = 1 + (v/b), u < 1\}$ by putting $\phi = 0$ for $u = 1$, $v = 0$ (see Figure 1).

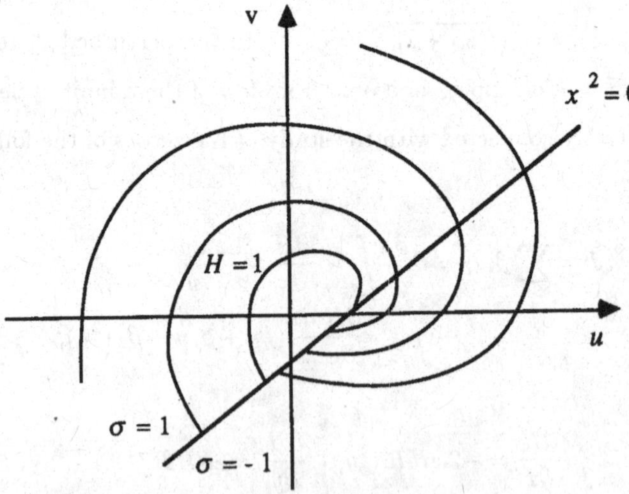

Figure 1

The equation $H = h$ corresponds to

$$Rh = e^{\phi/b}. \tag{8}$$

The domain $x^2 > 0$ corresponds to the domain $u > 1 + (v/b)$ for $\sigma = -1$ and $u < 1 + v/b$ for $\sigma = 1$ in Figure 1.

We see that if $\sigma = -1$ then the curve $H = 1$ consists of one point $u = 1$, $v = 0$ and $H \leq 1$ in the domain $u > 1 + v/b$. For $h < 1$ ϕ changes along $H = h$ from $\phi_1 \in (-\pi + \operatorname{arc} tgb, 0)$ to $\phi_2 \in (0, \operatorname{arc} tgb)$.

If $\sigma = 1$ then the curve $H = 1$ is tangent to the line $x^2 = 0$ (see Figure 1) and $\phi \in (0, \phi_* b))$. The domain of interest for us is given by $H \leq 1$. If $h < 1$ then ϕ changes from $\phi_1 \in (0, \operatorname{arc} tgb)$ to $\phi_2 \in (\pi + \operatorname{arc} tgb, \phi_*(b))$ along $H = h$.

The shapes of the curves $H = h$ at the (x, y)-plane are illustrated in Figure 2.

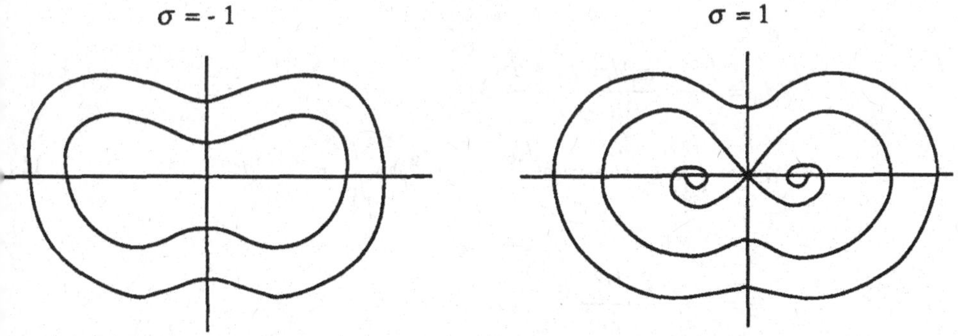

Figure 2

In order to get formulas simpler than (5) and (6) for the abelian integrals we shall make more changes. By (7) and (8) we have

$$\alpha x^2 = -\frac{\sigma}{h}(h + e^{\phi/b}(\frac{1}{b}\sin\phi - \cos\phi)) = -\frac{\sigma}{h}(h + e^{\phi/b}\frac{\sin(\phi-\phi_b)}{\sin\phi_b})$$

$$2\alpha x \cdot dx = -\frac{\sigma}{h}e^{\phi/b}\frac{\sin\phi}{\sin^2\phi_b}d\phi,$$

$$y = \frac{2\sigma}{bh}e^{\phi/b}\sin\phi$$

$$dy = \frac{2\sigma}{bh}e^{\phi/b}\frac{\sin(\phi-\phi_b)}{\sin\phi_b}d\phi, \quad \phi_b = \operatorname{arc} tgb. \tag{9}$$

Let us introduce a new variable

$$z = \sqrt{-aH(x,y)}x.$$

In the variables z and φ the Hamiltonian H takes the form

$$H = H^\sigma(z,\varphi) = \sigma z^2 - e^{\phi/b}\frac{\sin(\phi-\phi_b)}{\sin\phi_b}.$$

Then by (5)-(9)

$$J_{-1} = \frac{8}{b}|\alpha|^{3/2}\cdot\sigma\cdot h^{5/2}M_1,$$

$$M_{-1} = \int_{\phi_1}^{\phi_2} e^{-\phi/b}\frac{\sin(\phi-\phi_b)}{\sin\phi_b}\frac{d\phi}{z} = \frac{1}{2}\int_{\Gamma_h} e^{-\phi/b}\frac{\sin(\phi-\phi_b)}{\sin\phi_b}\frac{d\phi}{z}, \tag{10}$$

$$J_0 = \frac{-2\cdot|\alpha|^{1/2}\cdot\sigma\cdot h^{5/2}}{\sin^2\phi_b}M_0,$$

$$M_0 = \int_{\phi_1}^{\phi_2} e^{-\phi/b}\sin\phi\frac{d\phi}{z} = \sigma\sin^2\phi_b\int_{\Gamma_h} e^{-2\phi/b}dz, \tag{11}$$

$$J_1 = -\frac{4\cdot|\alpha|^{1/2}\cdot h^{3/2}}{b\cdot\sin^2\phi_b}M_1,$$

$$M_1 = \int_{\phi_1}^{\phi_2}\sin^2\phi\frac{d\phi}{z} = \sigma\sin^2\phi_b\int_{\Gamma_h} e^{-\phi/b}\sin\phi dz; \tag{12}$$

$$J_2 = \frac{4\cdot|\alpha|^{-1/2}\cdot h^{1/2}}{b\sin^2\phi_b}M_2, \quad M_2 = \int_{\phi_1}^{\phi_2}\sin^2\phi\cdot z\cdot d\phi. \tag{13}$$

Here $\Gamma_h = \Gamma_{h,\sigma}$ is the component of the curve $H^\sigma(z,\varphi) = h$ intersecting the line $\phi = 0$ for $\sigma = -1$ and $\phi = \pi$ for $\sigma = 1$, (see Figure 3).

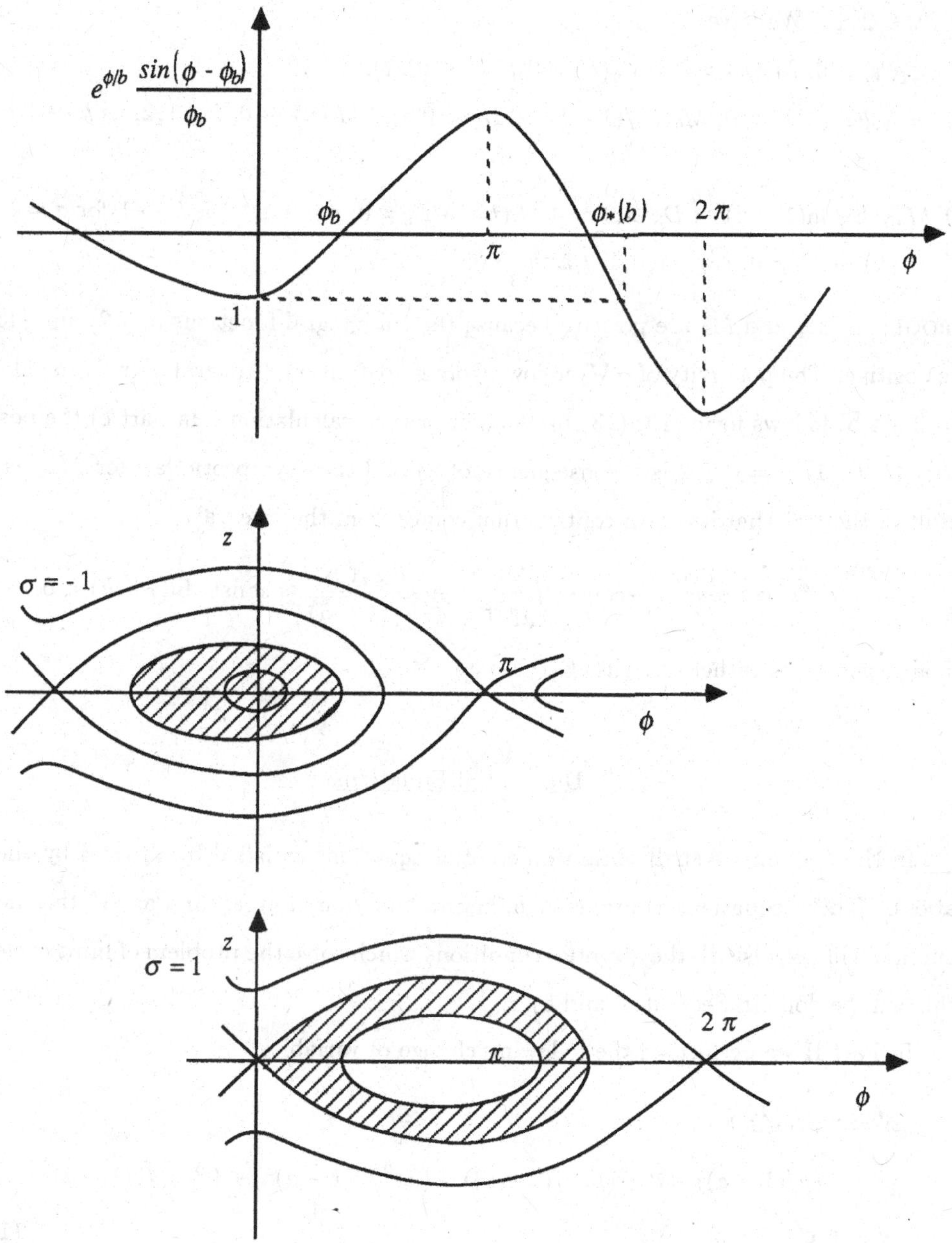

$e^{\phi/b} \frac{\sin(\phi - \phi_b)}{\phi_b}$
ϕ_b
π
$\phi_*(b)$
2π
ϕ
- 1
z
$\sigma = -1$
π
ϕ
z
$\sigma = 1$
π
2π
ϕ

Figure 3

LEMMA 2.1. *We have*

a) $M_1(h) > 0$, $M_2(h) > 0$, $\sigma M_0(h) > 0$ *for* $h \in (0,1)$;

b) $M_{-1}(h) \to D_1 > 0$, $M_0(h)/(1-h) \to D_2 < 0$ *and* $M_i(h) \to 0$, $i = 1,2$, *as* $h \to 1$ *for* $\sigma = -1$;

c) $M_{-1}(h)/\ln(1-h) \to D_3 > 0$ *and* $M_i(h) \to E_i \neq 0$, $i = 0,1,2$, *as* $h \to 1$ *for* $\sigma = 1$;

d) $M_i(h) \to F_i \neq 0$, $i = -1,0,1,2$ *as* $h \to 0$.

PROOF: a) M_1 and M_2 are positive because the subintegral functions in (12) and (13) are positive. The positivity of σM_0 follows from its definition, (integral over Γ_h in (11)).

Part b) follows form (10)-(13) by straightforward calculations. In part c) the positivity of $M_i(1)$, $i = 0,1,2$ is a consequence of a) and the asymptotic law for M_{-1} is a result of the fact that its main contribution comes from the integral

$$-\int_{\phi_1}^{\phi_2} e^{-\phi/b}\cos\phi \frac{d\phi}{z} \approx -\text{const} \int_{\sqrt{1-h}}^{1} \frac{d\phi}{\sqrt{\phi^2-(1-h)}} \approx \text{const}\cdot\ln(1-h) < 0.$$

Finally, part d) is rather obvious. ∎

3. Differential Equations

In this section we shall obtain differential equations satisfied by J_i (and by their ratios). These equations contain an unknown function, but as in Part II this new function will not disturb the essential conditions which solve the problem of limit cycles; (this will be done in Sections 4 and 5).

In Part II we performed the following change of variables

$$\begin{aligned} x' &= -x/(\sqrt{2}(1-a)) \\ y' &= -(y/(1-a)) - \sigma + (\beta_1 x/(1-a)) - (ax^2/2(1-a)^2) + (\beta_2 x^3/3(1-a)) \\ t' &= t/\sqrt{2}. \end{aligned} \tag{14}$$

Here a is the solution of the equation $2\alpha(1-a)^2 = a$. This change is real for $\alpha > -1/8$. We have obtained the system

$$\begin{aligned} \frac{dx'}{dt'} &= \sigma + ax'^2 + y' + \gamma_0 x'^{-1} + \gamma_1 x' + \gamma_2 x'^3 \\ \frac{dy'}{dt'} &= -2xy' - \delta_0 - \delta_1 x'^2 - \delta_2 x'^4 \end{aligned} \tag{15}$$

plus terms of higher order with respect to β_j. Obviously, the formula (14) in Part I does not contain the term $\gamma_0 x'^{-1}$ because the formula (4) there does not include the term $\beta_{-1}x^{-1}$, see (3). The constants in (15) are the following

$$\begin{aligned}
\gamma_0 &= \beta_{-1}/(\sqrt{2}(1-a)^2) \\
\gamma_1 &= \sqrt{2}\beta_1 \\
\gamma_2 &= 2\sqrt{2}\beta_2(1-a)^2/3 \\
\delta_0 &= \sqrt{2}a\beta_{-1}/(1-a)^2 + \sqrt{2}\beta_0/(1-a) \\
\delta_1 &= 2\sqrt{2}\beta_1 \\
\delta_2 &= 4\sqrt{2}(1-a)^2\beta_2/3
\end{aligned} \tag{16}$$

The system (15) for $\gamma_i = \delta_j = 0$ has the first integral

$$F = -a(a+1)y'^a(x'^2 + \frac{y'}{a+1} + \frac{\sigma}{a}). \tag{17}$$

As in Part I we introduce the integrals

$$\begin{aligned}
I_j &= \int_{F=c} y^{a-2}x^{2j+1}dy \\
K_j &= \int_{F=c} y^{a-1}x^{2j-1}dy \\
L_j &= \int_{F=c} y^{a-1}x^{2j}dx.
\end{aligned} \tag{18}$$

Then one has the following formula for the increment ΔF of the function F along an orbit of (15)

$$\Delta F = a(a+1)(\Sigma\gamma_j K_j + \Sigma\delta_j L_j). \tag{19}$$

The functions K_j and L_j may be represented as linear combinations of the functions I_j

$$L_j = \frac{1-a}{2j+1}I_j, \quad K_j = -\sigma I_{j-1} - \frac{(2j-1)a+2}{2j+1}I_j. \tag{20}$$

We recall also the relations between α, a, b and λ_j

$$\begin{aligned}
&\alpha = -(b^2+1)/8, \quad b > 0, \quad \lambda_{1,2} = \frac{1}{2}(1 \mp ib), \quad a = -\lambda_1/\lambda_2, \\
&1-a = 1/\lambda_2, \quad 1+a = ib/\lambda_2, \quad 2\alpha(1-a)^2 = a.
\end{aligned} \tag{21}$$

If $\alpha > -1/8$ then all the above transformations are real. Here we extend them to the case when $\alpha < -1/8$. Then a, $1-a$ and $1+a$ are complex

$$H = F^{i/(1-a)b} \tag{22}$$

(see (4)), and the corresponding functions I_j, K_j and L_j are complex. They are integrals along curves in $\mathbf{C}^2$, images of the curves $M = h$ in $\mathbf{R}^2$ under the transformation (14). The subintegral functions are as in (18).

From (22) we have

$$\Delta H = \frac{i}{b(1-a)} c^{\xi} \cdot \Delta F, \tag{23}$$

where $\xi = \frac{i}{(1-a)b} - 1$ and $\Delta H = \sum \beta_k J_k$. If we compare coefficients before β_k in both sides of (23) and perform some transformations, using (16), (18), (19) and (20), then we get

$$\begin{aligned} J_{-1} &= \sqrt{2}\alpha c^{\xi}(\sigma I_{-1} + a\frac{\alpha+1}{\alpha} I_0) \\ J_0 &= -\sqrt{2} \cdot a \cdot c^{\xi} \cdot I_0 \\ J_1 &= \sqrt{2} a c^{\xi}(\sigma I_0 + a I_1) \\ J_2 &= \frac{\sqrt{2}a^2}{3\alpha} c^{\xi}(\sigma I_1 + a I_2). \end{aligned} \tag{24}$$

In Part II (see equations (18) and (19) we have introduced the variables $Q_j = I_j/I_0$. They satisfy the equations

$$\begin{aligned} 2(a+1)c\frac{dQ_1}{dc} &= \frac{3\sigma}{a} + 2Q_1 - \frac{\sigma}{a}Q_1 Q_{-1} \\ 2(a+1)c\frac{dQ_2}{dc} &= \frac{5\sigma}{a}Q_1 + 4Q_2 - \frac{\sigma}{a}Q_2 Q_{-1} \end{aligned} \tag{25}$$

The functions Q_j are not real. We introduce new functions, which are real

$$\begin{aligned} R_1 &= -aQ_1 = J_1/J_0 + \sigma = \sigma(\frac{2M_1}{bhM_0} + 1) \\ R_2 &= -J_2/J_0 = \frac{2\sigma M_2}{bh^2 M_0} = \frac{\sigma a}{3} Q_1 + \frac{a^2}{3} Q_2 \\ T &= -\frac{\sigma}{a}(1-h)Q_{-1} = (1-h)[\frac{\alpha+1}{\alpha} - \frac{4b}{b^2+1}\frac{M_{-1}}{M_0}] \quad \text{for} \quad \sigma = -1 \\ T &= -\tau\sigma Q_{-1}/a = \tau[\frac{\alpha+1}{\alpha} - \frac{4b}{b^2+1}\frac{M_{-1}}{M_0}] \quad \text{for} \quad \sigma = 1, \end{aligned} \tag{26}$$

where

$$\tau = 1/(1 - \ln(1 - h)).$$

T is chosen in such a way that $T \to$ const as $h \to 1$; ($\tau = 0$ for $h = 1$ and $\tau = 1$ for $h = 0$). T is just the additional function mentioned at the beginning of this section.

Now, from (25) and (26) we obtain the differential equations for the variables h (or τ), R_1, R_2 and T. The graph of the function $h \to (R_1, R_2, T)(h)$ form one of the trajectories of this system.

LEMMA 3.1. *If $\sigma = -1$ then we have*

$$\begin{aligned} \dot{h} &= -2h(1-h) \\ \dot{R}_1 &= 3(1-h) + 2R_1(1-h) + R_1 T \\ \dot{R}_2 &= 1 - h + R_1(1-h) + 4R_2(1-h) + R_2 T \\ \dot{T} &= G(h) \end{aligned} \tag{27}$$

where $G(h) = 2h(1-h)(-dT/dh)$.

LEMMA 3.2. *If $\sigma = 1$ then we get the following equations*

$$\begin{aligned} \dot{\tau} &= 2\tau^2(1 - \exp(1 - 1/\tau)) \\ \dot{R}_1 &= -3\tau + 2R_1\tau + R_1 T \\ \dot{R}_2 &= \tau - R_1\tau + 4R_2\tau + R_2 T \\ \dot{T} &= \overline{G}(\tau) \end{aligned} \tag{28}$$

where $\overline{G}(\tau) = \dot{\tau}(\tau) \cdot dT/d\tau$.

Remark. The equations (27) and (28) should be obtained directly from the definitions of integrals J_i. The analysis presented above seems to be too complicated.

Denote

$$\Omega = \{(R_1, R_2)(h) : h \in (0, 1]\}. \tag{29}$$

The next lemma is analogous to Lemma 3.1 from Part I.

LEMMA 3.3. *The assertion of Theorem 1 is equivalent to the assertion that Ω is embedded, nonsingular and convex curve in* $\mathbf{R}^2$.

4. Proof of Theorem 1 for $\sigma = -1$

We start with the description of the behaviour of the functions R_i near the ends of their domains of definition.

LEMMA 4.1. *a)* $R_2 > 0$;

b) $R_i(h) \to 0$ *as* $h \to 1$ *and* $R_2 = R_1 + \frac{10}{27}R_1^2 + o(R_1^2)$ *as* $R_1 \to 0$;

c) $R_i \to \infty$ *as* $h \to 0$ *and* $R_2 = KR_1^2(1 + o(1))$ *as* $R_1 \to \infty$, $K > 0$.

PROOF: The points a) and c) follow from the formulas (11), (12), (13), (26) and from the points a) and b) of Lemma 2.1. To prove the point b) we can compute the integrals asymptotically. But it is better to use the Lemma 4.1 of Part II and the formulas (26). In Part I one has $Q_2 = (10/9)Q_1^2 + o(Q_1^2)$ as $Q_1 \to 0$. From these the point b) follows easily, since the formulas obtained turn out to be real. ∎

The proof of the following result is postponed to Section 7.

THEOREM 2. $dR_1/dh < 0$ *for* $h \in (0,1)$, $\sigma = -1$.

Now we are prepared to finish the proof of Theorem 1 in the case $\sigma = -1$. By Theorem 2 the curve Ω (29) can be represented in the form $\Omega : R_2 = W(R_1)$ and Theorem 1 is equivalent to the condition $W'' > 0$, (see Lemma 3.1).

Assume that W is not convex and $\overline{R} = (\overline{R}_1, \overline{R}_2)$ be the inflection point of W nearest to (0,0). Then

$$\lambda = W'(\overline{R}_1) > 0, \quad W''(\overline{R}_1) = 0, \quad W'''(\overline{R}_1) < 0 \tag{30}$$

or $W'''(\overline{R}_1) = 0$. In the latter case as in Part I we can find a trajectory of the system (27) close to our solution whose projection $\tilde{\Omega}$ onto the Q-plane has an inflection point $\tilde{R}_1$, satisfying $\tilde{W}'''(\tilde{R}_1) < 0$. See the proof of Lemma 6.1 in Part II. Therefore we assume that (30) holds.

We shall show that the last property in (30) is impossible.

Let $L = L_{\lambda,\omega} = \{(R_1, R_2) : R_2 = \lambda R_1 + \omega\}$ be the line tangent to Ω at the point $R \in \Omega$. The line L defines the hyperplane in $\mathbf{R}^4$

$$M = M_{\lambda,\omega} = \{(h, R_1, R_2, T) : R_2 = \lambda R_1 + \omega\}.$$

The vector field $X(P)$ (27) allows to associate with each point $P \in \mathbf{R}^4$ the line $Z = \{P + X(P)s : s \in R\}$ passing through P following the direction defined by $X(P)$.

We look at the points on the line $Z \subset M$, at which the vector field X is tangent to M. In other words we consider the function

$$\zeta(s) = (\dot{R}_2 - \lambda \dot{R}_1)\Big|_Z .$$

As in part I $\zeta(s)$ is a quadratic function

$$\zeta(s) = As^2 + Bs$$

and has two zeroes. Each zero corresponds to the different kind of contact of the vector field X (27) to M at these zeroes. At the inflection point, $B = 0$ and the character of tangency is determined by the sign of the coefficient A. For the situation described above, ($R = \overline{R}$, see (30)), $A < 0$ and the situation is illustrated in Figure 4, where the projections of the trajectories tangent to M are presented.

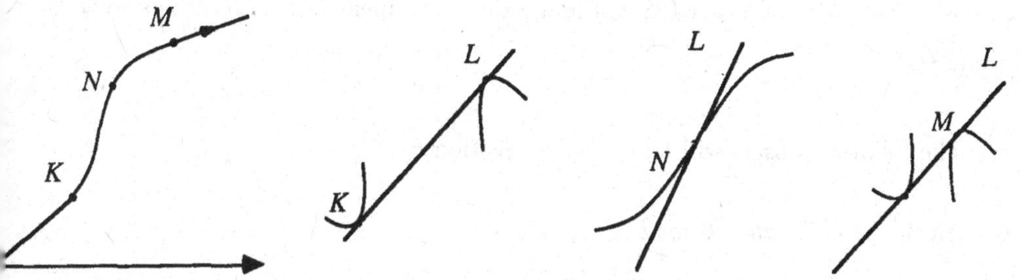

Figure 4

Now we can compute the coefficient A. We have (by (27))

$$(\dot{R}_2 - \lambda\dot{R}_1)\Big|_{R_2=\lambda R_1+\omega} = \omega T + (1-h)(4\omega+1-3\lambda) + R_1(1-h)(2\lambda+1). \quad (31)$$

The restriction of the function (31) to Z gives

$$\zeta(s) = -(2\lambda+1)\dot{R}_1\dot{h}s^2. \quad (32)$$

Because $\lambda > 0$, $\dot{R}_1 > 0$ and $\dot{h} < 0$, (see (27), and Theorem 2), the coefficient A is positive. This contradicts (30).

Theorem 1 in the case $\sigma = -1$ is complete.

5. Proof of Theorem 1 for $\sigma = 1$

As in the previous case we firstly describe the properties of the functions R_i near the ends of the interval (0,1). (Recall that we have introduced the new variable $\tau \in (0,1)$, $dh/d\tau < 0$.)

LEMMA 5.1. *a)* $R_i(h) > 0$ *for* $h \in (0,1]$ *and* $T|_{h=1} > 0$;
b) $\dot{R}_i\Big|_{h=1} > 0$ *and* $dR_2/dR_1|_{h=1} = R_2/R_1$;
c) sign $d^2R_2/dR_1^2|_{h=1} =$ *sign* $\Lambda|_{h=1}$, *where*

$$\Lambda = 2R_1R_2 + 3R_2 - R_1^2 + R_1; \quad (33)$$

d) $R_i \to \infty$ *as* $h \to 0$ $(\tau \to 1)$ *and* $R_2 = KR_1^2(1+o(1))$ *as* $R_1 \to \infty$, $K > 0$.

PROOF: Part a) follows from (10), (11), (12), (26) and the parts a) and c) of Lemma 2.1. Part d) is the same as in Lemma 4.1. Part b) follows from the fact that $\dot{R}_i = R_iT$, $T > 0$ for $h = 1$ $(\tau = 0)$, (see (28)). Finally the term linear in τ in $\dot{R}_2/\dot{R}_1$ is $(\Lambda/R_1^2T)\tau$, (see (28)).

The following result will be proved in Section 6.

THEOREM 3. $dR_1/dh < 0$ *for* $h \in (0,1)$, $\sigma = 1$.

We pass to our main goal, the proof of Theorem 1. We have $\dot{\tau} > 0$ and $\dot{R}_1 > 0$, (see (28) and Theorem 3).

As in the case $\sigma = -1$ we can show that if $\overline{R} = (\overline{R}_1, \overline{R}_2)$ is an inflection point of the curve $\Omega = \{(R_1, W(R_1))\}$ then

$$\operatorname{sign} W'''(\overline{R}_1) = \operatorname{sign} (2W'(\overline{R}_1) - 1), \tag{34}$$

(the corresponding coefficient A is equal to $(2\lambda - 1)\dot{R}_1\dot{\tau}$).

We have two possibilities.

I. $2W'(R_1)|_{h=1} \geq 1$,

II. $2W'(R_1)|_{h=1} < 1$.

Case I. By Lemma 5.1 b) and c) $\Lambda|_{h=1} > 0$ (or $W''|_{h=1} > 0$) and hence $W'''(\overline{R}_1) \leq 0$ and $W'(\overline{R}_1) > 1/2$ at the first inflection point $\overline{R}_1$ of W, (see Figure 5). This contradicts (34).

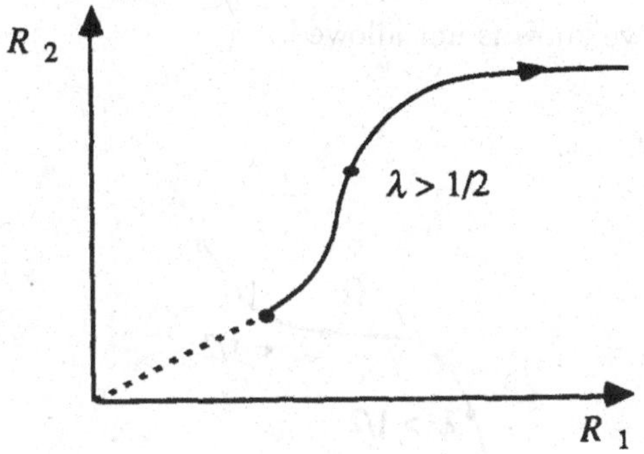

Figure 5

Case II. Here either

(i) $W''|_{h=1} \geq 0$ or

(ii) $W''|_{h=1} < 0$.

In the (i) the situation as in Figure 6 seems to be possible. But by (34) the values of W' at the both inflection points P and Q should be as is illustrated in Figure 6. Obviously, it is impossible. However, it may happen that the two inflection points coincide, $P = Q(\overline{R}_1, \overline{R}_2)$. Then $\lambda = W'(\overline{R}_1) = 1/2$. If $R_2 = \lambda R_1 + \omega$ is the line tangent to Ω at $P = Q$ and $Z = \{\tau + \dot{\tau}s, \overline{R} + s\dot{R}, T + s\dot{T}\}$ then

$$(\dot{R}_2 - \lambda\dot{R}_1)\Big|_{R_2=\lambda R_1+\omega} = \omega T + \tau(1 + 4\omega + 3\lambda) + (2\lambda - 1)R_1\tau$$

and $(\dot{R}_2 - \lambda\dot{R}_1)\Big| Z = 0$. Therefore we have two conditions

$$\tau - R_1\tau + 4R_2\tau + R_2T = \frac{1}{2}(-3\tau + 2R_1\tau + R_1T) \quad (\lambda = 1/2) \tag{35}$$

$$T \cdot \dot{\tau} = \overline{G}(\tau)\tau \quad (\zeta = 0). \tag{36}$$

We can perturb the initial condition for the system (28) in such a way that the condition (35) is achieved and the condition (36) fails, $\dot{T}\tau - T\dot{\tau} > 0$. We can make such perturbation arbitrary small. Then the restriction of the function $\dot{R}_2 - \lambda\dot{R}_1$ to the line Z gives the function $-Bs$, $B > 0$. Hence, by this and the smooth dependence of solutions of differential equations on initial conditions the perturbed trajectory is as shown in Figure 6, which we know is not allowed.

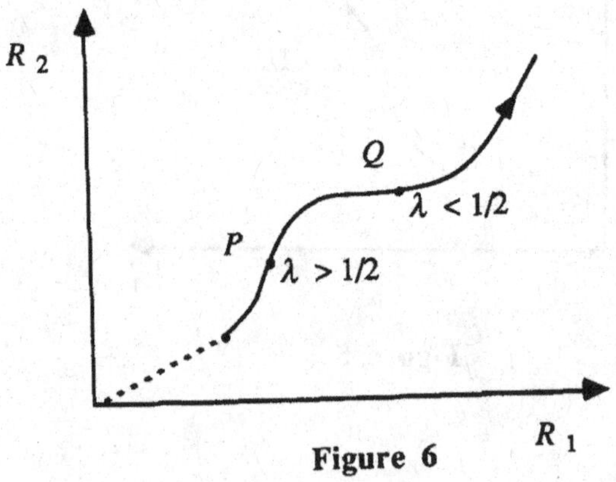

Figure 6

Consider case (ii). Then $W'(\overline{R}_1) < W'\big|_{h=1} < 1/2$ and $W'''(\overline{R}_1) < 0$ by (34). But from Figure 7 we see that it has to be $W'''(\overline{R}_1) \geq 0$. This contradiction completes the proof of Theorem 1 for $\sigma = 1$. ∎

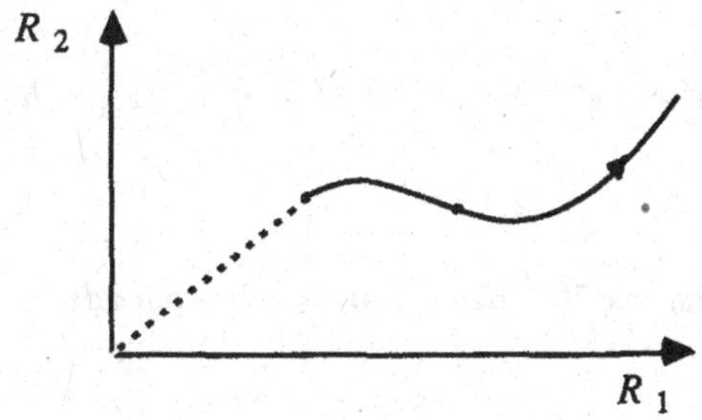

Figure 7

6. Proof of Theorem 3

6.1. A Formula for Derivative of the Integral. By (11), (12) and (26) the assertion of Theorem 3 states that

$$K = \frac{d}{dh}(hM_0/M_1) = M_0^{-2}\{M_0M_1 + h(M_0'M_1 - M_0M_1')\} > 0, \tag{37}$$

where M_0 and M_1 are given in (11) and (12) for $\sigma = 1$.

It turns out that the standard formulas for the derivatives of integrals, (for example $M_0' = (2/b)\sin^2\phi_b \int e^{-3\phi/b}\sin^{-1}\phi dz$), are not very useful in the proof of the inequality (37)). We use the formula first found by the author in [**28**].

Definition 6.1. We call the vector field X compatible with H iff

$$H_*X = Y \circ H, \tag{38}$$

where H is a function on $\mathbf{R}^2$ and $Y = Y(h)\cdot\partial/\partial h$ is some vector field in $\mathbf{R}^1$.

LEMMA 6.1. *Let (38) hold. Then*

$$\frac{d}{dh}\int_{H=h}\eta = Y^{-1}\int_{H=h}\mathcal{L}_X\eta, \tag{39}$$

where $\mathcal{L}$ denotes the Lie derivative.

In our situation we choose

$$H = H(z,\phi) = z^2 - (e^{\phi/b}\sin(\phi-\phi_b)/\sin\phi_b)),$$
$$X = z(\partial/\partial z) + f(\partial/\partial\phi), \tag{40}$$

where

$$f = 2 \cdot \sin^2 \phi_b \cdot e^{-\phi/b} \cdot \sin^{-1} \phi(z^2 - z_m^2), \quad z_m^2 = h + e^{\pi/b},$$

and the following forms

$$\eta_0 = e^{-2\phi/b} dz, \qquad \eta_1 = e^{-\phi/b} \sin \phi dz.$$

LEMMA 6.2. *a) X is compatible with H and $Y = 2z_m^2$;*
b) $\mathcal{L}_X \eta_0 = \eta_0 - (2/b) f \eta_0$;
c) $\mathcal{L}_X \eta_1 = \eta_1 - (z^2 - H) f \eta_0$.

PROOF: It follows from an immediate calculation. ∎

6.2. Reduction to Two Estimates. By (11), (12), (37) and Lemmas 6.1 and 6.2 we strive to estimate the expression

$$\frac{4M_0^2}{\sin^4 \phi_b} K = \int_0^{z_m} \int_0^{z_m} L(z, \tilde{z}) dz d\tilde{z}, \tag{41}$$

where

$$\begin{aligned} L(z, \tilde{z}) = {} & (e^{-\phi/b} \sin \phi \Big|_{\phi'}^{\phi})(z) \cdot (e^{-2\phi/b} \Big|_{\phi'}^{\phi})(\tilde{z}) \\ & + \frac{h}{2(e^{\pi/b} + h)} (-f e^{-2\phi/b} \Big|_{\phi'}^{\phi})(z) \\ & \cdot [(\frac{2}{b} e^{-\phi/b} \sin \phi - (z^2 - h) e^{-2\phi/b}) \Big|_{\phi'}^{\phi}](\tilde{z}), \end{aligned} \tag{42}$$

where $(g(\phi)|_{\phi'}^{\phi})(z) = g(\phi(z)) - g(\phi'(z))$ and $\phi(z) < \phi'(z)$ are the ϕ-components of the intersection of the curve $H = h$ with the line $z = $ const., (see Fig. 3).

PROPOSITION 6.1. *The function L is positive.*

Obviously, this result gives Theorem 3.

Now we give two estimates.

LEMMA 6.3. *We have the inequality*

$$(\frac{2}{b} e^{-\phi/b} \sin \phi \Big|_{\phi'}^{\phi})(\tilde{z}) \geq \frac{\tilde{z}^2 - h + 1}{1 + e^{-\pi/b}} (e^{-2\phi/b} \Big|_{\phi'}^{\phi})(\tilde{z}) \tag{43}$$

for $-1 < \tilde{z}^2 - h < e^{\pi/b}$, $\tilde{z} > 0$ and $h \in (0,1)$.

LEMMA 6.4. *We have the inequality*

$$(e^{-\phi/b}\sin\phi\Big|_{\phi'}^{\phi})(z) \geq \frac{z^2 - h}{2(1+e^{\pi/b})}(-fe^{-2\phi/b}\Big|_{\phi'}^{\phi})(z) \tag{44}$$

for $0 < h < z^2 < e^{\pi/b} + h$, $h \in (0,1)$.

PROOF OF PROPOSITION 6.1: a) Let $0 < z < \sqrt{h}$, $0 < \tilde{z} < z_m$. Then by (42), (43) and (44) we have

$$\begin{aligned} L(z,\tilde{z}) \geq & (e^{-2\phi/b}\Big|_{\phi'}^{\phi})(\tilde{z}) \cdot (e^{-\phi/b}\sin\phi\Big|_{\phi'}^{\phi})(z) \\ & + \frac{h}{2(e^{\pi/b}+h)}(-fe^{-2\phi/b}\Big|_{\phi'}^{\phi})(z) \\ & \cdot (e^{-2\phi/b}\Big|_{\phi'}^{\phi})(\tilde{z}) \cdot [\frac{\tilde{z}^2 - h + 1}{1+e^{-\pi/b}} - (z^2 - h)]. \end{aligned}$$

Now, because of (40) $f(\phi) < 0$, $f(\phi') > 0$, $(e^{-2\phi/b}\Big|_{\phi'}^{\phi}) > 0$, $(\sin\phi e^{-\phi/b}\Big|_{\phi'}^{\phi}) > 0$ and

$$\frac{\tilde{z}^2 - h + 1}{1+e^{-\pi/b}} - (z^2 - h) > 0 \quad \text{for} \quad 0 < z^2 < h, h \in (0,1).$$

From this the desired inequality follows.

b) Let $\sqrt{h} \leq z < z_m$, $0 < \tilde{z} < z_m$. Then

$$L(z,\tilde{z}) \geq (-fe^{-2\phi/b}\Big|_{\phi'}^{\phi})(z) \cdot (e^{-2\phi/b}\Big|_{\phi'}^{\phi})(\tilde{z}) \cdot \tilde{L}(z,\tilde{z}),$$

where

$$\tilde{L}(z,\tilde{z}) = \frac{z^2 - h}{2(e^{\pi/b}+1)} + \frac{h}{2(e^{\pi/b}+h)} \cdot [\frac{\tilde{z}^2 - h + 1}{1+e^{-\pi/b}} - z^2 + h] > \frac{\tilde{z}^2 - h + 1}{2(1+e^{\pi/b})^2} > 0$$

for $z^2 > h$. Hence L is positive in this case too. ∎

6.3. PROOF OF LEMMA 6.3. Observe that both sides of the inequality (43) are functions of one variable $t = \tilde{z}^2 - h \in (-1, e^{\pi/b})$. Before proving inequality (43) we notice that the inequality

$$(\frac{2}{b}e^{-\phi/b}\sin\phi\Big|_{\phi'}^{\phi}) - t \cdot (e^{-2\phi/b}\Big|_{\phi'}^{\phi}) > 0 \tag{45}$$

is obvious because the left hand side of (45) is

$$\frac{2}{b}\int_{\phi(t)}^{\phi'(t)} e^{-2\phi/b}\cdot[e^{\phi/b}\frac{\sin(\phi-\phi_b)}{\sin\phi_b}-t]d\phi,$$

where $e^{\phi/b}\sin(\phi-\phi_b)/\sin\phi_b > t$ for $\phi(t) < \phi < \phi'(t)$. The inequality (45) is not the best possible but it was our starting point and the better estimate (43) was found afterwards.

We show that

$$g_1(t) > 0 \quad \text{and} \quad g_2(t) < 0, \tag{46}$$

where $g_i(t) = g(\phi_i)$, $\phi_1 = \phi \in (0,\pi)$ and $\phi_2 = \phi' \in (\pi, \phi_*(b))$, (see Fig. 3), and

$$g(\phi) = \frac{2}{b}e^{-\phi/b}\cdot\sin\phi - \frac{t+1}{1+e^{-\pi/b}}\cdot(e^{-2\phi/b}-e^{-2\pi/b}),$$

here $t = t(\phi_i) = e^{\phi_i/b}\sin(\phi_i-\phi_b)/\sin\phi_b$.

LEMMA 6.5. *We have*

a) $g_1(-1) = g_i(e^{\pi/b}) = 0$;

b) $g_2(-1) = (2/b)\cdot\exp(-\phi_*/b)\cdot\sin(\phi_*(b)) < 0$.

PROOF: We should notice that $t(0) = t(\phi_*) = -1$ and $t(\pi) = e^{\pi/b}$. ∎

Now we calculate the derivatives of g_i using the formula $dt = e^{\phi/b}\cdot\sin\phi\cdot\sin^{-2}\phi_b d\phi$. We get

$$g_i' = \frac{2}{b}\cdot\frac{e^{-3\phi_i/b}}{\sin\phi_i}\cdot\sin^2\phi_b\cdot\frac{e^{\pi/b}-t}{e^{\pi/b}+1} - \frac{e^{-2\phi/b}-e^{-2\pi/b}}{e^{-\pi/b}+1}. \tag{47}$$

LEMMA 6.6. *We have*

a) $g_1'(t) \sim C/\sqrt{t+1}$, $C > 0$ *as* $t \to -1$;

b) $g_1'(t) \sim -C_1\sqrt{e^{\pi/b}-t}$, $C_1 > 0$ *as* $t \to e^{\pi/b}$;

c) $g_2'(t) \sim C_1\sqrt{e^{\pi/b}-t}$ *as* $t \to e^{\pi/b}$.

The proof uses the asymptotic formulas

$$t+1 \sim \phi_1^2/2\sin^2\phi_b, \quad e^{\pi/b}-t \sim e^{\pi/b}\cdot(\pi-\phi)^2/2\sin^2\phi_b. \quad ∎$$

The second derivatives of g_i are the following

$$g_i'' = \frac{2}{b}\cdot e^{-4\phi_i/b}\cdot\frac{\sin^2\phi_b\cdot\sin(\phi_i+\psi_b)}{\sin^3\phi_i\cdot(1+e^{\pi/b})}\cdot k_i(t),$$

where

$$k_i(t) = (e^{\pi/b} - 1) \cdot \frac{e^{\phi_i/b} \cdot \sin^2 \phi_i}{\sin(\phi_i + \psi_b)} - \frac{\sin^2 \phi_b}{\sin \psi_b}(e^{\pi/b} - t), \quad tg\psi_b = b/3.$$

The next lemma is obvious.

LEMMA 6.7. *$k_1(-1) < 0$ and $k_i(e^{\pi/b}) = 0$.*

LEMMA 6.8. *$k_i \neq 0$.*

PROOF: After some calculations we get

$$k_i' = (\alpha u^2 + \beta u + \gamma) \sin^2 \phi_b,$$

where

$$u = ctg(\phi_i + \psi_b), \quad \alpha = \frac{b}{\sqrt{b^2+9}}(e^{\pi/b} - 1), \quad \beta = \frac{2}{\sqrt{b^2+9}}(e^{\pi/b} - 1)$$

and

$$\gamma = ((3 + 2b^2)(e^{\pi/b} - 1)/b\sqrt{b^2+9}) + \sqrt{b^2+9}/b.$$

The discriminant of the above quadratic form is negative. ∎

Now we are ready to finish the proof of Lemma 6.3. By Lemmas 6.7 and 6.8 $k_i(t) \neq 0$ for $t \neq e^{\pi/b}$ and then $g_i'' \neq 0$. Hence $g_1'' < 0$ (see Lemma 6.7) and by Lemma 6.5 a) $g_1 > 0$. Moreover, by Lemma 6.6 c) $g_2'' > 0$ and $g_2 < 0$ by Lemma 6.5 b).

Lemma 6.3 is finished ∎

6.4. PROOF OF LEMMA 6.4. By (40) it is enough to show that

$$g_i(t) = (e^{\phi_i/b} \cdot \sin \phi_i)^2 - \sin^2 \phi_b \cdot t \cdot \frac{e^{\pi/b} - t}{e^{\pi/b} + 1} > 0, \quad t \in (0, e^{\pi/b}),$$

(see the beginning of Subsection 6.3). We have an obvious

LEMMA 6.9. *a) $g_i(0) > 0$;*

b) $g_i(e^{\pi/b}) = 0.$

Let us compute the derivative of g_i.

$$g_i' = 2e^{\phi_i/b} \cdot \sin(\phi_i + \phi_b) \cdot \sin\phi_b - \sin^2\phi_b \cdot \frac{e^{\pi/b} - 2t}{e^{\pi/b} + 1}.$$

LEMMA 6.10. *a)* $g_i'(e^{\pi/b}) < 0$;

b) $g_1'(0) > 0$ *for* $b \leq 2$.

PROOF: a) follows from an immediate calculation. Next

$$g_1'(0) = [(4e^{(\phi_b/b)}/\sqrt{b^2+1}) - (1 + e^{-\pi/b})^{-1}]\sin^2\phi_b,$$

where the function $(4/\sqrt{b^2+1}) - (1 + e^{-\pi/b})^{-1}$ is positive for $b \leq 2$. ∎

As before we find

$$g_i'' = [\frac{\sin(\phi_i + 2\phi_b)}{\sin\phi_i} + (1 + e^{\pi/b})^{-1}] \cdot 2 \cdot \sin^2\phi_b$$

and hence g_i'' is a linear function of $ctg\phi_i$ and has at most one zero, (because the range of values of ϕ_i is exactly π). Moreover $g_1''(t) \to -\infty$ and $g_2'' \to \infty$ as $t \to e^{\pi/b}$.

LEMMA 6.11. *If* $b \geq 2$ *then* $g_1''(0) < 0$.

PROOF: We have

$$g_1''(0) = 2\sin^2\phi_b[3 - \frac{4b^2}{b^2+1} + 1/(e^{\pi/b} + 1)].$$

But the function $3 - (4b^2/(b^2+1)) + (1/(e^{\pi/b}+1))$ is negative for $b \geq 2$. ∎

Consider g_1. If $b \geq 2$ then $g_1'' < 0$ and $g_1 > 0$ by Lemma 6.9. If $b \leq 2$ then $g_1'(0) > 0$ (see Lemma 6.10 b)) and even if g_1'' vanishes then, by Lemma 6.9, $g_1 > 0$.

Consider g_2. Then $g_2''(e^{\pi/b}) = \infty$ and even if g_2'' vanishes somewhere then by Lemmas 6.9 and 6.10 the function g_2 must be positive. ∎

Lemma 6.4 is complete.

7. Proof of Theorem 2

7.1. Preliminary Lemmas and General Ideas. By (26) we have to show that

$$\frac{d}{dh}(\sigma h M_0/M_1) > 0, \qquad \sigma = -1 \tag{48}$$

where M_0 and M_1 are given in (11) and (12). The integration in (11) and (12) runs over the curve $\Gamma_{h,\sigma}$ — the component of the h-level of the function

$$H^\sigma(\phi, z) = \sigma z^2 - e^{\phi/b} \cdot \sin(\phi - \phi_b)/\sin\phi_b$$

intersecting the line $\phi = 0$ for $\sigma = -1$ and $\phi = \pi$ for $\sigma = 1$.

Firstly we make some changes of variables in order to reduce the proof of the inequality (48) with $\sigma = -1$ to the analogous one with $\sigma = 1$ and $h < 0$. (For $\sigma = 1$ we have had $h \in (0,1)$.)

LEMMA 7.1. *The inequality (48) with $\sigma = -1$ and $h \in (0,1)$ is equivalent to the same inequality with $\sigma = 1$ and $h \in (-e^{\pi/b}, 0)$.*

PROOF: In the integrals in (38) we perform the following changes: $\phi' = \phi + \pi$, $h' = -he^{\pi/b}$ and $z' = -ze^{\pi/2b}$. Then $\Gamma_{h,-}$ transforms to $\Gamma_{h,+}$, (orientations are preserved), $M_0^-(h) = -M_0^+(h')e^{-5\pi/2b}$, $M_1^-(h) = M_1^+ e^{-3\pi/2b}$, $dM_0^-/dh = dM_0^+/dh' \cdot e^{-3\pi/2b}$ and $dM_1^-/dh = -dM_1^+/dh' e^{-\pi/2b}$. From this Lemma 7.1 follows. ∎

In what follows we assume that $\sigma = 1$, $h \in (-e^{\pi/b}, 0)$ and $\Gamma_h = \Gamma_{h,+}$.

Unfortunately, the proof of the inequality (48) presented in the previous section does not work in our case; it may be seen that asymptotically as $b \to 0$, the subintegral function L in (42) is not positive.

Here we use the formula (39) from Lemma 6.1 with

$$f = 2 \cdot \sin(\phi - \phi_b) \cdot \sin^{-1}\phi \cdot \sin\phi_b.$$

Notice that here the vector field $X = z \cdot (\partial/\partial z) + f(\partial/\partial\phi)$ is not finite, although the corresponding integrals remain bounded. We have

$$H_* X = 2h\partial/\partial h \tag{49}$$

and the other assertions of Lemma 6.2 remain unchanged.

Let us rewrite the inequality (48) for $\sigma = 1$

$$K = \int_0^{z_m}\int_0^{z_m} \tilde{L}(z,\tilde{z})dzd\tilde{z} > 0 \tag{50}$$

where

$$\begin{aligned}\tilde{L}(z,\tilde{z}) = &(e^{-2\phi/b}\Big|_{\phi'}^{\phi})(z)\cdot(e^{-\phi/b}\sin\phi\Big|_{\phi'}^{\phi})(\tilde{z})\\ &+(e^{-2\phi/b}\Big|_{\phi'}^{\phi})(z)\cdot(\tilde{z}^2-h)\cdot\sin\phi_b\cdot(e^{-2\phi/b}\frac{\sin(\phi-\phi_b)}{\sin\phi}\Big|_{\phi'}^{\phi})(\tilde{z})\\ &-\frac{2}{b}\sin\phi_b\cdot(e^{-2\phi/b}\frac{\sin(\phi-\phi_b)}{\sin\phi}\Big|_{\phi'}^{\phi})(z)\cdot(e^{-\phi/b}\sin\phi\Big|_{\phi'}^{\phi})(\tilde{z})\end{aligned} \tag{51}$$

almost exactly as in (41). We want to show that the function $\tilde{L}$ is positive. It is not difficult to check that it is not the case (at least for small b). What we do is to write

$$K = \frac{1}{2}\iint M(z,\tilde{z})dzd\tilde{z}, \tag{52}$$

where

$$M(z,\tilde{z}) = \tilde{L}(z,\tilde{z}) + \tilde{L}(\tilde{z},z). \tag{53}$$

The following result completes the proof of Theorem 2.

PROPOSITION 7.1. *The function M is positive.*

Remarks. 1. The above trick was used by Neishtadt **[18]** in his study of abelian integrals in the case of symmentry of order 4.

2. We could consider the function $L(z,\tilde{z})+L(\tilde{z},z)$ with L given in (42) but it is not positive. We could also replace the integration in (41) over $z's$ by the integration over $\phi's$. Obtained in such a way subintegral function is not positive. From this and from other results (see Part II, **[28]** and **[18]** for example) we observe that there are many formulas for derivative of the ratio of abelian integrals, but it seems that only one leads to the positive result. However there is no criterion for the choice of a good formula.

7.2. Reduction to Another Inequality. Here we start with the proof of Proposition 7.1. It relies upon many estimates and is rather complicated. Undoubtedly, there should exist a simpler proof but it resisted our efforts.

Obviously, the function $M(z,\tilde{z})$ depends only on $t = z^2 - h$ and $s = \tilde{z}^2 - h$, s, $t \in (-h, e^{\pi/b})$ but not on h. So it is enough to show the positivity of the function

$$N(\phi,\psi) = M(z,\tilde{z}), \tag{54}$$

where $\phi \in (\phi_b, \pi)$ and $\psi \in (\phi_b, \pi)$ are given by the formulas $t = e^{\phi/b} \cdot \sin(\phi - \phi_b)/\sin\phi_b$ and $s = e^{\psi/b} \cdot \sin(\psi - \phi_b)/\sin\phi_b$. Proposition 7.1 follows from the following two lemmas.

LEMMA 7.2. *$N(\phi,\phi) \to 0$ as $\phi \to \pi$.*

PROOF: Asymptotically $\tilde{\phi} = \pi - \phi \approx \phi' - \pi$, $e^{\pi/b} - t \approx e^{\pi/b}\tilde{\phi}^2/2\sin^2\phi_b$ and

$$(e^{-\phi/b}\sin\phi\big|_{\phi'}^{\phi})/(e^{-2\phi/b}\Big|_{\phi'}^{\phi}) \in (t, e^{\pi/b}) \tag{55}$$

(see the proof of Lemma 6.3). From this and from (51) the result follows. ∎

LEMMA 7.3. *$(\partial N/\partial\phi)(\phi,\psi) < 0$ for $\phi \le \psi$.*

The remaining part of this work is devoted to the proof of this lemma.

7.3. A Formula for $\partial N/\partial\phi$. In the calculations of the formula for $\partial N/\partial\phi$ we ought to remember that

$$\frac{d\phi'}{d\phi} = e^{(\phi-\phi')/b}\frac{\sin\phi}{\sin\phi'} < 0.$$

After some calculations we get

$$\frac{\partial N}{\partial\phi} = -\frac{2}{b}e^{-(2\phi+\psi)/b} \cdot \sin\psi \cdot (R+S), \tag{56}$$

where

$$\begin{aligned} R &= (1+\alpha)\cdot[1+\beta+(1-\frac{t}{s})\frac{(1+v(\psi))^2}{1+b^2}(1+\gamma)] = (1+\alpha)\cdot T,\\ S &= U\cdot W,\\ U &= 1+\beta-\frac{t}{s}\delta(1+v(\psi)),\\ W &= \frac{1}{1+b^2}[k(v(\phi))+k(v(\phi'))\alpha]. \end{aligned} \tag{57}$$

Moreover, here

$$v(\phi) = -b(ctg\phi)$$

$$\alpha = \alpha(\phi) = -e^{2(\phi-\phi')/b} \cdot \frac{\sin\phi}{\sin\phi'}$$

$$\beta = \beta(\psi) = -e^{(\psi-\psi')/b} \cdot \frac{\sin\psi'}{\sin\psi} > 0$$

$$\gamma = \gamma(\psi) = -e^{(\psi-\psi')/b} \cdot \frac{\sin\psi}{\sin\psi'} > 0$$

$$\delta = \delta(\psi) = \sqrt{1+b^2} \cdot (1 - e^{2(\psi-\psi')/b})/2 > 0$$

$$k(v) = v^2 - 2v + b^2 - 2. \tag{58}$$

LEMMA 7.4. a) $T > 0$; *b)* $U > 0$; *c)* $k(v(\phi')) > 0$ *for* $\phi' \in (\pi, \pi + \phi_b)$.

PROOF: a) is clear because $t \leq s$ by assumption. The inequality b) means that

$$(e^{-\psi/b}\sin\psi\Big|_{\psi'}^{\psi}) > t(e^{-2\psi/b}\Big|_{\psi'}^{\psi})$$

but it is a consequence of (55). To prove c) notice that $v(\phi') = -b\cdot ctg\phi' < -b\cdot ctg\phi_b = -1$ and hence $k(v(\phi')) > 1 + b^2 > 0$. Lemma 7.4 is complete. ∎

By Lemma 7.3 and by the inequalities (58) it is enough to show that

$$F(\phi,\psi) = (1+b^2)(1+\beta) + (1 - \frac{t}{s})(1+v(\psi))^2 + U \cdot k(v(\phi)) > 0. \tag{59}$$

LEMMA 7.5. *If* $b > 1$ *then* $F > 0$.

PROOF: Obviously, we have to consider only the case when $k(v(\phi)) < 0$. Then we get $F > (1+\beta)(v^2 - 2v + 2b^2 - 1)$. But the discriminant of $v^2 - 2v + 2b^2 - 1$ is $8(1-b^2) < 0$, for $b > 1$.

Next we proceed as follows. F is a function of two variables, but it is complicated. We shall show the series of inequalities for some functions of one variable, which together will lead to the inequality (59). We divide the domain of investigation into two parts:

I. $\phi_b < \phi < \pi - \phi_b$ or $-1 < v(\phi) < 1$,

II. $1 < v(\phi) < 1 + \sqrt{2(1-b^2)} = v_b$, ($v_b$ is the root of $v^2 - 2v + 2b^2 - 1$).

In the first case we show that F is decreasing in ϕ and then we prove its positivity on the diagonal. In the second case we add some negative term to F and then proceed as in the first case with a new function.

7.4 The case I. $-1 < v < 1$.

LEMMA 7.6. *$\partial F/\partial \phi < 0$.*

PROOF: Since the function $k(v)$ is decreasing for $v < 1$ by (57) and (59) it is enough to show that

$$\frac{(1+v(\psi))^2}{s} + \frac{\partial U}{\partial t} k(v(\phi)) = \frac{1+v(\psi)}{s}[1 + v(\psi) - \delta k(v(\phi))] > 0.$$

Obviously it will be sufficient if we prove that

$$P(v) = 1 + v - \delta k(v) > 0 \quad \text{for} \quad v \in (-1, 1).$$

Because $\delta < \frac{1}{2}\sqrt{1+b^2} < \frac{1}{2} + \frac{1}{4}b^2$ (see (58)), then

$$P(v) > \frac{2+b^2}{4}v^2 - \frac{1}{2}b^2 v + b^2/4 > (v - b^2)^2/4 \geq 0.$$

From this Lemma 7.6 follows. ∎

LEMMA 7.7. *$F(\phi, \phi) > 0$ if $-1 < v(\phi) < 1$.*

PROOF: Firstly we give certain estimates on the functions β and δ, which we shall prove later.

LEMMA 7.8. *a) $\beta < \frac{3}{10}(b^2 + v^2)$;*
b) $\delta > (\frac{1}{2} + \frac{b^2}{5})(1 - \frac{b^2}{10} - \frac{v^2}{10})$.

Using Lemma 7.8, (57) and (59) for $\phi = \psi$ we estimate $F(\phi, \phi)$ by an expression, which is a polynomial of b^2 with coefficients depending on v. We have

$$F(\phi, \phi) > Q_0(v) + b^2 Q_1(v) + b^4 Q_2(v) + b^6 Q_3(v).$$

Calculations give

$$Q_0 = \frac{v^2}{20}(22 - 26v + 5v^2 + v^3) > 0 \qquad \text{for} \quad |v| < 1,$$

$$Q_1 = \frac{1}{100}(150 - 50v + 106v^2 - 18v^3 - 2v^4 + 2v^5) > 0 \qquad \text{for} \quad |v| < 1,$$

$$Q_2 = \frac{1}{100}(41 - 23v + 4v^3) > 0 \qquad \text{for} \quad |v| < 1,$$

$$Q_3 = \frac{1+v}{50} > 0.$$

Therefore it suffices to prove Lemma 7.8. ∎

PROOF OF LEMMA 7.8: Denote

$$x = (\pi - \phi)/b, \qquad y = (\phi' - \pi)/b.$$

Then $x > (\sin(bx)/b) = 1/\sqrt{v^2 + b^2}$ and hence

$$e^{-x} \le (\frac{2}{e})^2(v^2 + b^2) < 0.55(v^2 + b^2) \quad \text{and} \quad e^{-2x} \le \frac{v^2 + b^2}{e^2} < \frac{v^2 + b^2}{7}.$$

Therefore we should show that, (see (58))

$$e^{-y}\frac{\sin(by)}{\sin(bx)} < \frac{3}{10}/0.55 \approx 0.54 \tag{60}$$

and

$$e^{-2y} < 2/3 \tag{61}$$

(see (58)). Here y is the solution of the equation

$$\begin{aligned} f(b, y) = e^y[\cos(by) - \frac{\sin(by)}{b}] &= e^{-x}[\cos(bx) + \frac{\sin(bx)}{b}] \\ &= e^{-x} \cdot \frac{v+1}{\sqrt{v^2 + b^2}}. \end{aligned} \tag{62}$$

We divide the problem into two parts:

(i) $-1 < v < 0$ or $\phi < \pi/2$ and

(ii) $0 < v < 1$ or $\pi/2 < \phi < \pi - \phi_b$.

In the case (i) $y(v) \ge y(0)$ for $v < 0$. Then

$$f(b, y) < \frac{1}{b}e^{-\pi/2b} \le (\frac{2}{e})^2 \cdot (\frac{2}{\pi})^2 \cdot b < \frac{1}{4}.$$

On the other hand $y < \phi_b/b < 1$, (since $\sin\phi_b = b/\sqrt{b^2+1} < \sin b$), $e^y > 1+y$, $\cos(by) > 1 - 2by/\pi$ (for $by < \pi/2$) and $\sin(by) < by$. Therefore

$$f(b,1/2) > \frac{3}{2}(1 - \frac{b}{\pi} - \frac{1}{2}) > \frac{1}{4} \quad \text{for} \quad 0 < b < 1.$$

From this and from the fact that $f(b,y)$ is decreasing in y and the right hand side of (62) is increasing in v, we find

$$y > y|_{v=0} > 1/2. \tag{63}$$

Hence

$$e^{-x-y} < e^{-1/2}e^{-\pi/2b} \leq e^{-1/2}(\frac{2}{e})^2(\frac{2}{\pi})^2 b^2 < \frac{1}{5}(v^2+b^2). \tag{64}$$

Now consider the function

$$g(\phi) = |\sin\phi'|/\sin\phi.$$

LEMMA 7.9. *a) $g(\phi_b) = g(\pi) = 1$;*
b) $g(\phi) < 1$ for $\phi \in (\phi_b, \pi)$;
c) $g'(\phi) > 0$ for $\phi \in (\pi/2, \pi)$.

PROOF: a) is rather obvious. The other assertions follow from the fact that $\phi + \phi'$ increases as ϕ increases. This follows from Figure 3 and can be easily proved analytically. ∎

Now, from Lemma 7.9, (63) and (64) the inequalities (60) and (61) follow.

Consider the case (ii). By Lemma 7.9 c) we have to prove the inequality (60) for $\phi = \pi - \phi_b$. In order to do it we should estimate $y|_{v=1}$ from (62). We have

$$e^{-x}(v+1)/\sqrt{v^2+b^2} \leq 2/e < 4/5.$$

We solve the equation

$$f(b,y) = 4/5, \qquad y = y(b).$$

LEMMA 7.10. *$dy/db < 0$.*

PROOF: $dy/db = -(\partial f/\partial b)/(\partial f/\partial y)$, where $df/dy = -(b+(1/b))\cdot e^y \cdot \sin(by) < 0$ and

$$df/db = \frac{e^y}{b^2}[(1-b^2y)\sin(by) - by\cos(by)] = \frac{e^y}{b^2}g(b,\alpha),$$

$0 < \alpha = by < \phi_b < b$. We have $g(b,0) = 0$, $g(b,\phi_b) < 0$ and $\partial g/\partial \alpha < 0$. Therefore $g < 0$ and the Lemma is proved. ∎

Due to this result we compute

$$y(1) \approx 0.38, \qquad y(0) \approx 0.52$$

and hence

$$e^{-y} < 0.69 < \sqrt{2/3},$$
$$\frac{\sin(by)}{\sin(bx)} < \left.\frac{\sin(by)}{\sin(bx)}\right|_{v=1} < y\sqrt{b^2+1} < 0.74.$$

From this we obtain the inequalities (60) and (61).

7.5. The Case II. $1 < v < v_b$. Here instead of the function F (see (59)) we consider the function

$$G(\phi,\psi) = F(\phi,\psi) - U \cdot 2\sqrt{2(1-b^2)}(v(\phi)-1) < F(\phi,\psi) \tag{65}$$

and prove its positivity.

LEMMA 7.11. *$\partial G/\partial \phi < 0$ for $\phi < \psi$.*

PROOF:

$$G = (1+b^2)(1+\beta) + (1-\frac{t}{s})(1+v(\psi))^2$$
$$+ [1+\beta - \frac{t}{s}\delta(1+v(\psi))](v^2 - 2vv_b + b^2 - 2 + 2\sqrt{2(1-b^2)}),$$

where the polynomial

$$r(v) = v^2 - 2vv_b + b^2 - 2 + 2\sqrt{2(1-b^2)} \tag{66}$$

is decreasing for $v < v_b$. Therefore we want to show the inequality

$$1 + v + \delta r(v) > 0. \tag{67}$$

Now $\delta < \sqrt{1+b^2}/2$ and hence

$$1+v+\delta r(v) > \frac{\sqrt{1+b^2}}{2}v^2 - (\sqrt{b^2+1}v_b - 1)v + 1 + \frac{b^2}{2}\sqrt{1+b^2} - \sqrt{1+b^2} + \sqrt{2(1-b^4)}.$$

The discriminant of the quadratic polynomial in the right hand side of this inequality is

$$-b^4/2 - 2\sqrt{2(1-b^4)} < 0.$$

Lemma 7.11 is complete. ∎

The next lemma completes the proof of Theorem 2.

LEMMA 7.12. *$G(\phi,\phi) > 0$ for $1 < v(\phi) < v_b$.*

PROOF: Firstly we estimate β and δ.

LEMMA 7.13. *a) $\beta < (v^2+b^2)/4(v+1)$;*
b) $\delta > \frac{1}{2}(1+\frac{2b^2}{5})[1 - 0.64\frac{v^2+b^2}{(v+1)}(1+b^2)]$.

PROOF: By (58) and (62)

$$\beta = e^{-x-y}\frac{\sin(by)}{\sin(bx)} = e^{-x}\sqrt{v^2+b^2}\cdot\frac{\sin(by)}{b}\cdot e^{x}\cdot\frac{\sqrt{v^2+b^2}}{v+1}(\cos(by) - \frac{\sin(by)}{b}) \le C(v^2+b^2)/(v+1),$$

where

$$C = \sup_y \frac{\sin(by)}{b}\sqrt{1+b^2}\frac{\sin(\phi_b - by)}{b} = 1/(2(1+\sqrt{1+b^2})) \le 1/4.$$

To prove b) we have to estimate

$$e^{-x-y} = \frac{\sqrt{v^2+b^2}}{v+1}\cdot\sqrt{1+b^2}\cdot\frac{\sin(\phi_b - by)}{b} < \frac{\sqrt{v^2+b^2}\cdot\sqrt{1+b^2}}{2}\cdot(1-y). \qquad (68)$$

The expression $1-y$ takes its maximal value for $v = v_b = 1+\sqrt{2(1-b^2)}$. We fix $v = v_0 = 1+\sqrt{2}$ (maximal possible) and solve the equation (62). The smallest possible y is for $b = 1$ (see Lemma 7.10). Calculations shows that $y > 0.2$. From this, from (68) and from the inequality $\sqrt{2+b^2} > 1+\frac{2}{5}b^2$ the assertion of Lemma 7.13 follows. ∎

We pass to the proof of Lemma 7.12. We denote $u = (v-1)/\sqrt{2}$, $0 \le u \le 1$ and use the estimates $\sqrt{1-b^2} < 1 - b^2/2$ and $(2+0.8b^2)(1+b^2) < 2 + 3.6b^2$. Then from (65) we get

$$\begin{aligned}4(v+1)G(\phi,\phi) > {} & (1+b^2)(9+6\sqrt{2}u+2u^2+b^2)\\ & + [9+6\sqrt{2}u+2u^2+b^2-(2+0.8b^2)(2u^2+4\sqrt{2}u+4)\\ & + 0.64(2+3.6b^2)(2u^2+2\sqrt{2}u+1+b^2)]\\ & \cdot (2u^2-4u-3+b^2(1+2u)).\end{aligned}$$

We treat this expression as a polynomial of b^2 i.e.,

$$4(v+1)G > \Sigma Q_i(u) b^{2i}.$$

Here

$$Q_0 > 2.1 - 3.2u + 1.6u^2 - 0.8u^3 + 1.2u^4$$

and it is easy to check that this polynomial is positive for $u \in [0,1]$. Next

$$Q_1 > 8.9 + 5.8u - 5.8u^2 - 6.1u^3 + 4.8u^4$$

and is positive too, but

$$Q_2 > -4 - 4.9u + 8.8u^2 + 4.8u^3$$

is not positive. However if $Q_2 < 0$ then

$$Q_1 + b^2 Q_2 > Q_1 + Q_2 > 0 \quad \text{for} \quad u \in [0,1].$$

Finally

$$Q_3 > 2(1+2u) > 0.$$

This completes the proof of Lemma 7.12 and of Theorem 2. ∎

References

1. Arnold, V. I., Geometrical Methods in the Theory of Ordinary Differential Equations, Springer-Verlag, New York, (1983).
2. Bogdanov, R. I., Versal Deformations of Singular Point of Vector Field on the Plane in the Zero Eigenvalues Case, in: Proc. of Petrovski Sem., 2, 37-65 (1976).
3. Basikin, A. D., Kuznietzov, Yu. A., Khibnik, A. I., Bifurcational Diagrams of Dynamical Systems on the Plane, Computer Center, Puschino, (1985), (in Russian).
4. Bateman, H., Erdelyi, A., Higher Transcendental Functions, V. 1, McGraw Hill Book Comp., New York, (1953).
5. Berezovskaya, F. S., Khibnik, A. I., in: Methods of Qualitative Theory of Differential Equations, Gorki Univers., Gorki, (1985), (in Russian).
6. Carr, J., Chow, S.-N., Hale, J., Abelian Integrals and Bifurcation Theory, J. Diff. Equat., 59, 413-436, (1985).
7. Carr, J., Van Gils, S.A., Sanders, J., Nonresonant Bifurcation with Symmetry, SIAM J. Math. Anal., 18 (3), 579-591, (1987).
8. Chow, S. - N., Li, C., Wang D., Uniqueness of Periodic Orbits in Some Vector Fields with Codimension two Singularities, J. Diff. Equat., 77 (2), 231-253, 1989.
9. Dumortier, F. Singularities of Vector Fields in the Plane Jour. Diff. Eq. 23 (1), 53–106, (1977).
10. Dumortier, F., Roussarie, R., Sotomayor, J., Generic 3-parameter Families of Vector Fields on the Plane. Unfolding a Singularity with Nilpotent Linear Part. The Cusp Case of Codimension 3, Ergodic Theory and Dynamical Systems, 7, 375–413, (1987).
11. Dumortier, F., Roussarie, R., Sotomayor, J., Generic 3-parameter Families of Planar Vector Fields. Unfolding of Saddle, Focus and Elliptic Singularities with Nilpotent Linear Parts; this volume.
12. Ecalle, Y., Martinet, J., Moussou, R., Ramis, J. – P., Non-accumulation des cycles-limites, R. C. Acad. Sc. Paris, 304 (I), Nr. 14, 375-377, 431-434, (1987).
13. van Gils, S. A., A note on "Abelian Integrals and Bifurcation Theory", J. Diff. Equat., 59, 437-439, (1985).
14. Iliashenko, Yu. S., On Zeroes of Special Abelian Integrals in Real Domain, Funct. Anal. Appl., 11 (4), 301-311, (1977).
15. Iliashenko, Yu. S., Uspiekhi Mat. Nauk, (to appear).
16. Khovansky, A. G., Real Analytic Manifolds with Finitness Properties and Complex Abelian Integrals, Funct. Anal. Appl., 18, 119-128, (1984).
17 Medved, M., The Unfolding of a Germ of Vector Field in the Plane with a Singularity of Codimension 3., Czech. Math. Journal, 35 (110), 1-41, (1985).

18. Neishtadt, A. I., Bifurcations of Phase Portrait of Certain System of Equations Arising in the Problem of Loss of Stability of Selfoscillations near Resonance 1:4, Prikl. Mat. Mech., 42(5), 896-907, (1978).

19. Petrov, G. S., Elliptic Integrals and their Non-oscillation, Funct. Anal. Appl., 20 (1), 37–40, (1986).

20. Petrov G.S., Complex Zeroes of an Elliptic Integral, Funct. Anal. Appl., 21 (3), 247-248, (1987).

21. Petrov G.S., Chebyshev Property of Elliptic Integrals, Funct. Anal. Appl., 22 (1), 72-73, (1988).

22 Petrov G.S., Complex Zeroes of an Elliptic Integral, Funct. Anal. Appl., 23 (2), 88-89, (1989), (Russian).

23 Roussarie R., Deformations Generiques des Cusps, Asterisque, 150– 151, 151–184, (1987).

24. Rousseau C., Zoladek J., Zeroes of Complete Elliptic Integrals in Real Domain, J. Diff. Equat., (to appear).

25. Varchenko, A. N., Estimate of the Number of Zeroes of Abelian Integrals Depending on Parameters and Limit Cycles, Funct. Anal. Appl., 18 (2), 98-108, (1984).

26. Yakovlenko, S. Yu., On the Real Zeroes of the Class of Abelian Integrals Arising in Bifurcation Theory, in: Methods of Qualitative Theory of Diff. Equat., Gorki Univers., Gorki, 175-185, (1984), (in Russian).

27. Ye Y. Q. and others, "Theory of Limit Cycles", Translation of Mathematical Monographs, AMS, V. 66, (1984).

28. Zoladek, H., On Versality of Certain Family of Symmetric Vector Fields on the Plane, Math. Sborn., 48 (2), 463-492, (1984),

29. Zoladek H., Bifurcations of Certain Family of Planov Vector Fields Tangent to Axes, J. Diff. Equat., 67 (1), 1-55, (1987).

30. Zoladek H., Abelian Integrals in Non-symmetric Perturbation of Symmetric Hamiltonian Vector Field, Adv. Appl. Math., (to appear).

31. Zhang Z. - F., van Gils, S.A., Drachman, B., Abelian Integrals for Quadratic Vector Fields, J. Reine Angew. Math., 382, 165-180, (1987).

INDEX

Vol. 1341: M. Dauge, Elliptic Boundary Value Problems on Corner Domains. VIII, 259 pages. 1988.

Vol. 1342: J.C. Alexander (Ed.), Dynamical Systems. Proceedings, 1986–87. VIII, 726 pages. 1988.

Vol. 1343: H. Ulrich, Fixed Point Theory of Parametrized Equivariant Maps. VII, 147 pages. 1988.

Vol. 1344: J. Král, J. Lukes, J. Netuka, J. Vesely´ (Eds.), Potential Theory – Surveys and Problems. Proceedings, 1987. VIII, 271 pages. 1988.

Vol. 1345: X. Gomez-Mont, J. Seade, A. Verjovski (Eds.), Holomorphic Dynamics. Proceedings, 1986. VII, 321 pages. 1988.

Vol. 1346: O.Ya. Viro (Ed.), Topology and Geometry – Rohlin Seminar. XI, 581 pages. 1988.

Vol. 1347: C. Preston, Iterates of Piecewise Monotone Mappings on an Interval. V, 166 pages. 1988.

Vol. 1348: F. Borceux (Ed.), Categorical Algebra and its Applications. Proceedings, 1987. VIII, 375 pages. 1988.

Vol. 1349: E. Novak, Deterministic and Stochastic Error Bounds in Numerical Analysis. V, 113 pages. 1988.

Vol. 1350: U. Koschorke (Ed.), Differential Topology Proceedings, 1987. VI, 269 pages. 1988.

Vol. 1351: I. Laine, S. Rickman, T. Sorvali (Eds.), Complex Analysis, Joensuu 1987. Proceedings. XV, 378 pages. 1988.

Vol. 1352: L.L. Avramov, K.B. Tchakerian (Eds.), Algebra – Some Current Trends. Proceedings. 1986. IX, 240 Seiten. 1988.

Vol. 1353: R.S. Palais, Ch.-l. Teng, Critical Point Theory and Submanifold Geometry. X, 272 pages. 1988.

Vol. 1354: A. Gómez, F. Guerra, M.A. Jiménez, G. López (Eds.), Approximation and Optimization. Proceedings, 1987. VI, 280 pages. 1988.

Vol. 1355: J. Bokowski, B. Sturmfels, Computational Synthetic Geometry. V, 168 pages. 1989.

Vol. 1356: H. Volkmer, Multiparameter Eigenvalue Problems and Expansion Theorems. VI, 157 pages. 1988.

Vol. 1357: S. Hildebrandt, R. Leis (Eds.), Partial Differential Equations and Calculus of Variations. VI, 423 pages. 1988.

Vol. 1358: D. Mumford, The Red Book of Varieties and Schemes. V, 309 pages. 1988.

Vol. 1359: P. Eymard, J.-P. Pier (Eds.) Harmonic Analysis. Proceedings, 1987. VIII, 287 pages. 1988.

Vol. 1360: G. Anderson, C. Greengard (Eds.), Vortex Methods. Proceedings, 1987. V, 141 pages. 1988.

Vol. 1361: T. tom Dieck (Ed.), Algebraic Topology and Transformation Groups. Proceedings. 1987. VI, 298 pages. 1988.

Vol. 1362: P. Diaconis, D. Elworthy, H. Föllmer, E. Nelson, G.C. Papanicolaou, S.R.S. Varadhan. École d´ Été de Probabilités de Saint-Flour XV–XVII. 1985–87 Editor: P.L. Hennequin. V, 459 pages. 1988.

Vol. 1363: P.G. Casazza, T.J. Shura, Tsirelson's Space. VIII, 204 pages. 1988.

Vol. 1364: R.R. Phelps, Convex Functions, Monotone Operators and Differentiability. IX, 115 pages. 1989.

Vol. 1365: M. Giaquinta (Ed.), Topics in Calculus of Variations. Seminar, 1987. X, 196 pages. 1989.

Vol. 1366: N. Levitt, Grassmannians and Gauss Maps in PL-Topology. V, 203 pages. 1989.

Vol. 1367: M. Knebusch, Weakly Semialgebraic Spaces. XX, 376 pages. 1989.

Vol. 1368: R. Hübl, Traces of Differential Forms and Hochschild Homology. III, 111 pages. 1989.

Vol. 1369: B. Jiang, Ch.-K. Peng, Z. Hou (Eds.), Differential Geometry and Topology. Proceedings, 1986–87. VI, 366 pages. 1989.

Vol. 1370: G. Carlsson, R.L. Cohen, H.R. Miller, D.C. Ravenel (Eds.), Algebraic Topology. Proceedings, 1986. IX, 456 pages. 1989.

Vol. 1371: S. Glaz, Commutative Coherent Rings. XI, 347 pages. 1989.

Vol. 1372: J. Azéma, P.A. Meyer, M. Yor (Eds.), Séminaire de Probabilités XXIII. Proceedings. IV, 583 pages. 1989.

Vol. 1373: G. Benkart, J.M. Osborn (Eds.), Lie Algebras. Madison 1987. Proceedings. V, 145 pages. 1989.

Vol. 1374: R.C. Kirby, The Topology of 4-Manifolds. VI, 108 pages. 1989.

Vol. 1375: K. Kawakubo (Ed.), Transformation Groups. Proceedings, 1987. VIII, 394 pages, 1989.

Vol. 1376: J. Lindenstrauss, V.D. Milman (Eds.), Geometric Aspects of Functional Analysis. Seminar (GAFA) 1987–88. VII, 288 pages. 1989.

Vol. 1377: J.F. Pierce, Singularity Theory, Rod Theory, and Symmetry-Breaking Loads. IV, 177 pages. 1989.

Vol. 1378: R.S. Rumely, Capacity Theory on Algebraic Curves. III, 437 pages. 1989.

Vol. 1379: H. Heyer (Ed.), Probability Measures on Groups IX. Proceedings, 1988. VIII, 437 pages. 1989.

Vol. 1380: H.P. Schlickewei, E. Wirsing (Eds.), Number Theory, Ulm 1987. Proceedings. V, 266 pages. 1989.

Vol. 1381: J.-O. Strömberg, A. Torchinsky, Weighted Hardy Spaces. V, 193 pages. 1989.

Vol. 1382: H. Reiter, Metaplectic Groups and Segal Algebras. XI, 128 pages. 1989.

Vol. 1383: D.V. Chudnovsky, G.V. Chudnovsky, H. Cohn, M.B. Nathanson (Eds.), Number Theory. New York 1985–88. Seminar. V, 256 pages. 1989.

Vol. 1384: J. Garcia-Cuerva (Ed.), Harmonic Analysis and Partial Differential Equations. Proceedings, 1987. VII, 213 pages. 1989.

Vol. 1385: A.M. Anile, Y. Choquet-Bruhat (Eds.), Relativistic Fluid Dynamics. Seminar, 1987. V, 308 pages. 1989.

Vol. 1386: A. Bellen, C.W. Gear, E. Russo (Eds.), Numerical Methods for Ordinary Differential Equations. Proceedings, 1987. VII, 136 pages. 1989.

Vol. 1387: M. Petkovi´c, Iterative Methods for Simultaneous Inclusion of Polynomial Zeros. X, 263 pages. 1989.

Vol. 1388: J. Shinoda, T.A. Slaman, T. Tugué (Eds.), Mathematical Logic and Applications. Proceedings, 1987. V, 223 pages. 1989.

Vol. 1000: Second Edition. H. Hopf, Differential Geometry in the Large. VII, 184 pages. 1989.

Vol. 1389: E. Ballico, C. Ciliberto (Eds.), Algebraic Curves and Projective Geometry. Proceedings, 1988. V, 288 pages. 1989.

Vol. 1390: G. Da Prato, L. Tubaro (Eds.), Stochastic Partial Differential Equations and Applications II. Proceedings, 1988. VI, 258 pages. 1989.

Vol. 1391: S. Cambanis, A. Weron (Eds.), Probability Theory on Vector Spaces IV. Proceedings, 1987. VIII, 424 pages. 1989.

Vol. 1392: R. Silhol, Real Algebraic Surfaces. X, 215 pages. 1989.

Vol. 1393: N. Bouleau, D. Feyel, F. Hirsch, G. Mokobodzki (Eds.), Séminaire de Théorie du Potentiel Paris, No. 9. Proceedings. VI, 265 pages. 1989.

Vol. 1394: T.L. Gill, W.W. Zachary (Eds.), Nonlinear Semigroups, Partial Differential Equations and Attractors. Proceedings, 1987. IX, 233 pages. 1989.

Vol. 1395: K. Alladi (Ed.), Number Theory, Madras 1987. Proceedings. VII, 234 pages. 1989.

Vol. 1396: L. Accardi, W. von Waldenfels (Eds.), Quantum Probability and Applications IV. Proceedings. 1987. VI, 355 pages. 1989.

Vol. 1397: P.R. Turner (Ed.), Numerical Analysis and Parallel Processing. Seminar. 1987. VI, 264 pages. 1989.

Vol. 1398: A.C. Kim, B.H. Neumann (Eds.), Groups – Korea 1988. Proceedings. V, 189 pages. 1989.

Vol. 1399: W.-P. Barth, H. Lange (Eds.), Arithmetic of Complex Manifolds. Proceedings, 1988. V, 171 pages. 1989.

Vol. 1400: U. Jannsen, Mixed Motives and Algebraic K-Theory. XIII, 246 pages. 1990.

Vol. 1401: J. Steprans, S. Watson (Eds.). Set Theory and its Applications. Proceedings. 1987. V, 227 pages. 1989.

Vol. 1402: C. Carasso, P. Charrier, B. Hanouzet, J.-L. Joly (Eds.), Nonlinear Hyperbolic Problems. Proceedings. 1988. V, 249 pages. 1989.

Vol. 1403: B. Simeone (Ed.), Combinatorial Optimization. Seminar, 1986. V, 314 pages. 1989.

Vol. 1404: M.-P. Malliavin (Ed.), Séminaire d´Algèbre Paul Dubreil et Marie-Paul Malliavin. Proceedings, 1987–1988. IV, 410 pages. 1989.

Vol. 1405: S. Dolecki (Ed.), Optimization. Proceedings, 1988. V, 223 pages. 1989. Vol. 1406: L. Jacobsen (Ed.), Analytic Theory of Continued Fractions III. Proceedings. 1988. VI, 142 pages. 1989.

Vol. 1407: W. Pohlers, Proof Theory. VI, 213 pages. 1989.

Vol. 1408: W. Lück, Transformation Groups and Algebraic K-Theory. XII, 443 pages. 1989.

Vol. 1409: E. Hairer, Ch. Lubich, M. Roche. The Numerical Solution of Differential-Algebraic Systems by Runge-Kutta Methods. VII, 139 pages. 1989.

Vol. 1410: F.J. Carreras, O. Gil-Medrano, A.M. Naveira (Eds.), Differential Geometry. Proceedings, 1988. V, 308 pages. 1989.

Vol. 1411: B. Jiang (Ed.), Topological Fixed Point Theory and Applications. Proceedings. 1988. VI, 203 pages. 1989.

Vol. 1412: V.V. Kalashnikov, V.M. Zolotarev (Eds.). Stability Problems for Stochastic Models. Proceedings, 1987. X, 380 pages. 1989.

Vol. 1413: S. Wright, Uniqueness of the Injective III_1 Factor. III, 108 pages. 1989.

Vol. 1414: E. Ramirez de Arellano (Ed.), Algebraic Geometry and Complex Analysis. Proceedings, 1987. VI, 180 pages. 1989.

Vol. 1415: M. Langevin, M. Waldschmidt (Eds.), Cinquante Ans de Polynômes. Fifty Years of Polynomials. Proceedings, 1988. IX, 235 pages.1990.

Vol. 1416: C. Albert (Ed.), Géométrie Symplectique et Mécanique. Proceedings. 1988. V, 289 pages. 1990.

Vol. 1417: A.J. Sommese, A. Biancofiore, E.L. Livorni (Eds.), Algebraic Geometry. Proceedings. 1988. V, 320 pages. 1990.

Vol. 1418: M. Mimura (Ed.), Homotopy Theory and Related Topics. Proceedings, 1988. V, 241 pages. 1990.

Vol. 1419: P.S. Bullen, P.Y. Lee, J.L. Mawhin, P. Muldowney, W.F. Pfeffer (Eds.), New Integrals. Proceedings, 1988. V, 202 pages. 1990.

Vol. 1420: M. Galbiati, A. Tognoli (Eds.), Real Analytic Geometry. Proceedings, 1988. IV, 366 pages. 1990.

Vol. 1421: H.A. Biagioni, A Nonlinear Theory of Generalized Functions, XII, 214 pages. 1990.

Vol. 1422: V. Villani (Ed.), Complex Geometry and Analysis. Proceedings, 1988. V, 109 pages. 1990.

Vol. 1423: S.O. Kochman, Stable Homotopy Groups of Spheres: A Computer-Assisted Approach. VIII, 330 pages. 1990.

Vol. 1424: F.E. Burstall, J.H. Rawnsley, Twistor Theory for Riemannian Symmetric Spaces. III, 112 pages. 1990.

Vol. 1425: R.A. Piccinini (Ed.), Groups of Self-Equivalences and Related Topics. Proceedings. 1988. V, 214 pages. 1990.

Vol. 1426: J. Azéma, P.A. Meyer, M. Yor (Eds.), Séminaire de Probabilités XXIV, 1988/89. V, 490 pages. 1990.

Vol. 1427: A. Ancona, D. Geman, N. Ikeda, École d'Eté de Probabilités de Saint Flour XVIII, 1988. Ed.: P.L. Hennequin. VII, 330 pages. 1990.

Vol. 1428: K. Erdmann, Blocks of Tame Representation Type and Related Algebras. XV, 312 pages. 1990.

Vol. 1429: S. Homer, A. Nerode, R.A. Platek, G.E. Sacks, A. Scedrov, Logic and Computer Science. Seminar, 1988. Editor: P. Odifreddi. V, 162 pages. 1990.

Vol. 1430: W. Bruns, A. Simis (Eds.), Commutative Algebra. Proceedings. 1988. V, 160 pages. 1990.

Vol. 1431: J.G. Heywood, K. Masuda, R. Rautmann, V.A. Solonnikov (Eds.), The Navier-Stokes Equations – Theory and Numerical Methods. Proceedings. 1988. VII, 238 pages. 1990.

Vol. 1432: K. Ambos-Spies, G.H. Müller, G.E. Sacks (Eds.), Recursion Theory Week. Proceedings. 1989. VI, 393 pages. 1990.

Vol. 1433: S. Lang, W. Cherry, Topics in Nevanlinna Theory. II, 174 pages.1990.

Vol. 1434: K. Nagasaka, E. Fouvry (Eds.), Analytic Number Theory. Proceedings. 1988. VI, 218 pages. 1990.

Vol. 1435: St. Ruscheweyh, E.B. Saff, L.C. Salinas, R.S. Varga (Eds.), Computational Methods and Function Theory. Proceedings, 1989. VI, 211 pages. 1990.

Vol. 1436: S. Xambó-Descamps (Ed.), Enumerative Geometry. Proceedings, 1987. V, 303 pages. 1990.

Vol. 1437: H. Inassaridze (Ed.), K-theory and Homological Algebra. Seminar, 1987–88. V, 313 pages. 1990.

Vol. 1438: P.G. Lemarié (Ed.) Les Ondelettes en 1989. Seminar. IV, 212 pages. 1990.

Vol. 1439: E. Bujalance, J.J. Etayo, J.M. Gamboa, G. Gromadzki. Automorphism Groups of Compact Bordered Klein Surfaces: A Combinatorial Approach. XIII, 201 pages. 1990.

Vol. 1440: P. Latiolais (Ed.), Topology and Combinatorial Groups Theory. Seminar, 1985–1988. VI, 207 pages. 1990.

Vol. 1441: M. Coornaert, T. Delzant, A. Papadopoulos. Géométrie et théorie des groupes. X, 165 pages. 1990.

Vol. 1442: L. Accardi, M. von Waldenfels (Eds.), Quantum Probability and Applications V. Proceedings, 1988. VI, 413 pages. 1990.

Vol. 1443: K.H. Dovermann, R. Schultz, Equivariant Surgery Theories and Their Periodicity Properties. VI, 227 pages. 1990.

Vol. 1444: H. Korezlioglu, A.S. Ustunel (Eds.), Stochastic Analysis and Related Topics VI. Proceedings, 1988. V, 268 pages. 1990.

Vol. 1445: F. Schulz, Regularity Theory for Quasilinear Elliptic Systems and – Monge Ampère Equations in Two Dimensions. XV, 123 pages. 1990.

Vol. 1446: Methods of Nonconvex Analysis. Seminar, 1989. Editor: A. Cellina. V, 206 pages. 1990.

Vol. 1447: J.-G. Labesse, J. Schwermer (Eds), Cohomology of Arithmetic Groups and Automorphic Forms. Proceedings, 1989. V, 358 pages. 1990.

Vol. 1448: S.K. Jain, S.R. López-Permouth (Eds.), Non-Commutative Ring Theory. Proceedings, 1989. V, 166 pages. 1990.

Vol. 1449: W. Odyniec, G. Lewicki, Minimal Projections in Banach Spaces. VIII, 168 pages. 1990.

Vol. 1450: H. Fujita, T. Ikebe, S.T. Kuroda (Eds.), Functional-Analytic Methods for Partial Differential Equations. Proceedings, 1989. VII, 252 pages. 1990.

Vol. 1451: L. Alvarez-Gaumé, E. Arbarello, C. De Concini, N.J. Hitchin, Global Geometry and Mathematical Physics. Montecatini Terme 1988. Seminar. Editors: M. Francaviglia, F. Gherardelli. IX, 197 pages. 1990.

Vol. 1452: E. Hlawka, R.F. Tichy (Eds.), Number-Theoretic Analysis. Seminar, 1988–89. V, 220 pages. 1990.

Vol. 1453: Yu.G. Borisovich, Yu.E. Gliklikh (Eds.), Global Analysis – Studies and Applications IV. V, 320 pages. 1990.

Vol. 1454: F. Baldassari, S. Bosch, B. Dwork (Eds.), p-adic Analysis. Proceedings, 1989. V, 382 pages. 1990.

Vol. 1455: J.-P. Françoise, R. Roussarie (Eds.), Bifurcations of Planar Vector Fields. Proceedings, 1989. VI, 396 pages. 1990.

Vol. 1456: L.G. Kovács (Ed.), Groups – Canberra 1989. Proceedings. XII, 198 pages. 1990.

Vol. 1457: O. Axelsson, L.Yu. Kolotilina (Eds.), Preconditioned Conjugate Gradient Methods. Proceedings, 1989. V, 196 pages. 1990.

Vol. 1458: R. Schaaf, Global Solution Branches of Two Point Boundary Value Problems. XIX, 141 pages. 1990.

Vol. 1459: D. Tiba, Optimal Control of Nonsmooth Distributed Parameter Systems. VII, 159 pages. 1990.

Vol. 1460: G. Toscani, V. Boffi, S. Rionero (Eds.), Mathematical Aspects of Fluid Plasma Dynamics. Proceedings, 1988. V, 221 pages. 1991.

Vol. 1461: R. Gorenflo, S. Vessella, Abel Integral Equations. VII, 215 pages. 1991.

Vol. 1462: D. Mond, J. Montaldi (Eds.), Singularity Theory and its Applications. Warwick 1989, Part I. VIII, 405 pages. 1991.

Vol. 1463: R. Roberts, I. Stewart (Eds.), Singularity Theory and its Applications. Warwick 1989, Part II. VIII, 322 pages. 1991.

Vol. 1464: D. L. Burkholder, E. Pardoux, A. Sznitman, Ecole d'Eté de Probabilités de Saint- Flour XIX-1989. Editor: P. L. Hennequin. VI, 256 pages. 1991.

Vol. 1465: G. David, Wavelets and Singular Integrals on Curves and Surfaces. X, 107 pages. 1991.

Vol. 1466: W. Banaszczyk, Additive Subgroups of Topological Vector Spaces. VII, 178 pages. 1991.

Vol. 1467: W. M. Schmidt, Diophantine Approximations and Diophantine Equations. VIII, 217 pages. 1991.

Vol. 1468: J. Noguchi, T. Ohsawa (Eds.), Prospects in Complex Geometry. Proceedings, 1989. VII, 421 pages. 1991.

Vol. 1469: J. Lindenstrauss, V. D. Milman (Eds.), Geometric Aspects of Functional Analysis. Seminar 1989-90. XI, 191 pages. 1991.

Vol. 1470: E. Odell, H. Rosenthal (Eds.), Functional Analysis. Proceedings, 1987-89. VII, 199 pages. 1991.

Vol. 1471: A. A. Panchishkin, Non-Archimedean L-Functions of Siegel and Hilbert Modular Forms. VII, 157 pages. 1991.

Vol. 1472: T. T. Nielsen, Bose Algebras: The Complex and Real Wave Representations. V, 132 pages. 1991.

Vol. 1473: Y. Hino, S. Murakami, T. Naito, Functional Differential Equations with Infinite Delay. X, 317 pages. 1991.

Vol. 1474: S. Jackowski, B. Oliver, K. Pawałowski (Eds.), Algebraic Topology, Poznań 1989. Proceedings. VIII, 397 pages. 1991.

Vol. 1475: S. Busenberg, M. Martelli (Eds.), Delay Differential Equations and Dynamical Systems. Proceedings, 1990. VIII, 249 pages. 1991.

Vol. 1476: M. Bekkali, Topics in Set Theory. VII, 120 pages. 1991.

Vol. 1477: R. Jajte, Strong Limit Theorems in Noncommutative L_2-Spaces. X, 113 pages. 1991.

Vol. 1478: M.-P. Malliavin (Ed.), Topics in Invariant Theory. Seminar 1989-1990. VI, 272 pages. 1991.

Vol. 1479: S. Bloch, I. Dolgachev, W. Fulton (Eds.), Algebraic Geometry. Proceedings, 1989. VII, 300 pages. 1991.

Vol 1480: F. Dumortier, R. Roussarie, J. Sotomayor, H. Żołądek, Bifurcations of Planar Vector Fields: Nilpotent Singularities and Abelian Integrals. VIII, 226 pages. 1991.